T0326356

System Parameter Identification

System Parameter Identification
Information Criteria and Algorithms

Badong Chen

Institute of Artificial Intelligence and Robotics (IAIR),
Xi'an Jiaotong University, Xi'an, China

Yu Zhu

Department of Mechanical Engineering,
Tsinghua University, Beijing, China

Jinchun Hu

Department of Mechanical Engineering,
Tsinghua University, Beijing, China

Jose C. Principe

Department of Electrical and Computer Engineering,
University of Florida, Gainesville, FL, USA

ELSEVIER

AMSTERDAM • BOSTON • HEIDELBERG • LONDON • NEW YORK • OXFORD
PARIS • SAN DIEGO • SAN FRANCISCO • SINGAPORE • SYDNEY • TOKYO

Elsevier
32 Jamestown Road, London NW1 7BY, UK
225 Wyman Street, Waltham, MA 02451, USA

First edition 2013

© 2013 Tsinghua University Press Ltd. Published by Elsevier Inc. All rights reserved.

No part of this publication may be reproduced or transmitted in any form or by any means, electronic or mechanical, including photocopying, recording, or any information storage and retrieval system, without permission in writing from the publisher. Details on how to seek permission, further information about the Publisher's permissions policies and our arrangement with organizations such as the Copyright Clearance Center and the Copyright Licensing Agency, can be found at our website: www.elsevier.com/permissions

This book and the individual contributions contained in it are protected under copyright by the Publisher (other than as may be noted herein).

Notices
Knowledge and best practice in this field are constantly changing. As new research and experience broaden our understanding, changes in research methods, professional practices, or medical treatment may become necessary.

Practitioners and researchers must always rely on their own experience and knowledge in evaluating and using any information, methods, compounds, or experiments described herein. In using such information or methods they should be mindful of their own safety and the safety of others, including parties for whom they have a professional responsibility.

To the fullest extent of the law, neither the Publisher nor the authors, contributors, or editors, assume any liability for any injury and/or damage to persons or property as a matter of products liability, negligence or otherwise, or from any use or operation of any methods, products, instructions, or ideas contained in the material herein.

British Library Cataloguing-in-Publication Data
A catalogue record for this book is available from the British Library

Library of Congress Cataloging-in-Publication Data
A catalog record for this book is available from the Library of Congress

ISBN: 978-0-12-404574-3

For information on all Elsevier publications
visit our website at store.elsevier.com

This book has been manufactured using Print On Demand technology. Each copy is produced to order and is limited to black ink. The online version of this book will show color figures where appropriate.

Working together
to grow libraries in
developing countries

ELSEVIER Book Aid International

www.elsevier.com • www.bookaid.org

Contents

About the Authors

Badong Chen received the B.S. and M.S. degrees in control theory and engineering from Chongqing University, in 1997 and 2003, respectively, and the Ph.D. degree in computer science and technology from Tsinghua University in 2008. He was a post-doctoral researcher with Tsinghua University from 2008 to 2010 and a post-doctoral associate at the University of Florida Computational NeuroEngineering Laboratory during the period October 2010 to September 2012. He is currently a professor at the Institute of Artificial Intelligence and Robotics, Xi'an Jiaotong University. His research interests are in system identification and control, information theory, machine learning, and their applications in cognition and neuroscience.

Yu Zhu received the B.S. degree in radio electronics in 1983 from Beijing Normal University, and the M.S. degree in computer applications in 1993 and the Ph.D. degree in mechanical design and theory in 2001, both from China University of Mining and Technology. He is currently a professor with the Department of Mechanical Engineering, Tsinghua University. His research field mainly covers IC manufacturing equipment development strategy, ultra-precision air/maglev stage machinery design theory and technology, ultra-precision measurement theory and technology, and precision motion control theory and technology. He has more than 140 research papers and 100 (48 awarded) invention patents.

Jinchun Hu, associate professor, born in 1972, graduated from Nanjing University of Science and Technology. He received the B.E. and Ph.D. degrees in control science and engineering in 1994 and 1998, respectively. Currently, he works at the Department of Mechanical Engineering, Tsinghua University. His current research interests include modern control theory and control systems, ultra-precision measurement principles and methods, micro/nano motion control system analysis and realization, special driver technology and device for precision motion systems, and super-precision measurement and control.

Jose C. Principe is a distinguished professor of electrical and computer engineering and biomedical engineering at the University of Florida where he teaches advanced signal processing, machine learning, and artificial neural networks modeling. He is BellSouth Professor and the founding director of the University of Florida Computational NeuroEngineering Laboratory. His primary research interests are in advanced signal processing with information theoretic criteria (entropy and mutual information) and adaptive models in reproducing kernel Hilbert spaces, and the application of these advanced algorithms to brain machine interfaces. He is a Fellow of the IEEE, ABME, and AIBME. He is the past editor in chief of the IEEE Transactions on Biomedical Engineering, past chair of the Technical Committee on Neural Networks of the IEEE Signal Processing Society, and past President of the International Neural Network Society. He received the IEEE EMBS Career Award and the IEEE Neural Network Pioneer Award. He has more than 600 publications and 30 patents (awarded or filed).

Preface

System identification is a common method for building the mathematical model of a physical plant, which is widely utilized in practical engineering situations. In general, the system identification consists of three key elements, i.e., the data, the model, and the criterion. The goal of identification is then to choose one from a set of candidate *models* to fit the *data* best according to a certain *criterion*. The criterion function is a key factor in system identification, which evaluates the consistency of the model to the actual plant and is, in general, an objective function for developing the identification algorithms. The identification performances, such as the convergence speed, steady-state accuracy, robustness, and the computational complexity, are directly related to the criterion function.

Well-known identification criteria mainly include the least squares (LS) criterion, minimum mean square error (MMSE) criterion, and the maximum likelihood (ML) criterion. These criteria provide successful engineering solutions to most practical problems, and are still prevalent today in system identification. However, they have some shortcomings that limit their general use. For example, the LS and MMSE only consider the second-order moment of the error, and the identification performance would become worse when data are non-Gaussian distributed (e.g., with multimodal, heavy-tail, or finite range). The ML criterion requires the knowledge of the conditional probability density function of the observed samples, which is not available in many practical situations. In addition, the computational complexity of the ML estimation is usually high. Thus, selecting a new criterion beyond second-order statistics and likelihood function is attractive in problems of system identification.

In recent years, criteria based on information theoretic descriptors of entropy and dissimilarity (divergence, mutual information) have attracted lots of attentions and become an emerging area of study in signal processing and machine learning domains. Information theoretic criteria (or briefly, information criteria) can capture higher order statistics and information content of signals rather than simply their energy. Many studies suggest that information criteria do not suffer from the limitation of Gaussian assumption and can improve performance in many realistic scenarios. Combined with nonparametric estimators of entropy and divergence, many adaptive identification algorithms have been developed, including the practical gradient-based batch or recursive algorithms, fixed-point algorithms (no step-size), or other advanced search algorithms. Although many elegant results and techniques have been developed over the past few years, till now there is no book devoted to a systematic study of system identification under information theoretic criteria. The primary focus of this book is to provide an overview of these developments, with emphasis on the nonparametric estimators of information criteria and gradient-based identification algorithms. Most of the contents of this book originally appeared in the recent papers of the authors.

The book is divided into six chapters: the first chapter is the introduction to the information theoretic criteria and the state-of-the-art techniques; the second chapter presents the definitions and properties of several important information measures; the third chapter gives an overview of information theoretic approaches to parameter estimation; the fourth chapter discusses system identification under minimum error entropy criterion; the fifth chapter focuses on the minimum information divergence criteria; and the sixth chapter changes the focus to the mutual information-based criteria.

It is worth noting that the information criteria can be used not only for system parameter identification but also for system structure identification (e.g., model selection). The Akaike's information criterion (AIC) and the minimum description length (MDL) are two famous information criteria for model selection. There have been several books on AIC and MDL, and in this book we don't discuss them in detail. Although most of the methods in this book are developed particularly for system parameter identification, the basic principles behind them are universal. Some of the methods with little modification can be applied to blind source separation, independent component analysis, time series prediction, classification and pattern recognition.

This book will be of interest to graduates, professionals, and researchers who are interested in improving the performance of traditional identification algorithms and in exploring new approaches to system identification, and also to those who are interested in adaptive filtering, neural networks, kernel methods, and online machine learning.

The authors are grateful to the National Natural Science Foundation of China and the National Basic Research Program of China (973 Program), which have funded this book. We are also grateful to the Elsevier for their patience with us over the past year we worked on this book. We also acknowledge the support and encouragement from our colleagues and friends.

Xi'an
P.R. China
March 2013

Symbols and Abbreviations

The main symbols and abbreviations used throughout the text are listed as follows.

$\lvert . \rvert$	absolute value of a real number
$\lVert . \rVert$	Euclidean norm of a vector
$\langle .,. \rangle$	inner product
$\mathbb{I}(.)$	indicator function
$E[.]$	expectation value of a random variable
$f'(x)$	first-order derivative of the function $f(x)$
$f''(x)$	second-order derivative of the function $f(x)$
$\nabla_x f(x)$	gradient of the function $f(x)$ with respect to x
$\mathbf{sign}(.)$	sign function
$\Gamma(.)$	Gamma function
$(.)^T$	vector or matrix transposition
I	identity matrix
A^{-1}	inverse of matrix A
$\det A$	determinant of matrix A
$\mathbf{Tr}A$	trace of matrix A
$\mathbf{rank}A$	rank of matrix A
$\log(.)$	natural logarithm function
z^{-1}	unit delay operator
$\mathbb{R}$	real number space
$\mathbb{R}^n$	n-dimensional real Euclidean space
$\rho(X,Y)$	correlation coefficient between random variables X and Y
$\mathbf{Var}[X]$	variance of random variable X
$\mathbf{Pr}[A]$	probability of event A
$\mathcal{N}(\mu, \Sigma)$	Gaussian distribution with mean vector μ and covariance matrix Σ
$\mathrm{U}[a,b]$	uniform distribution over interval $[a,b]$
$\chi^2(k)$	chi-squared distribution with k degree of freedom
$H(X)$	Shannon entropy of random variable X
$H_\psi(X)$	ϕ-entropy of random variable X
$H_\alpha(X)$	α-order Renyi entropy of random variable X
$V_\alpha(X)$	α-order information potential of random variable X
$S_\alpha(X)$	survival information potential of random variable X
$H_\Delta(X)$	Δ-entropy of discrete random variable X
$I(X;Y)$	mutual information between random variables X and Y
$D_{\mathrm{KL}}(X\|Y)$	KL-divergence between random variables X and Y
$D_\phi(X\|Y)$	ϕ-divergence between random variables X and Y
$J_{\mathbf{F}}$	Fisher information matrix
$\bar{J}_{\mathbf{F}}$	Fisher information rate matrix

$p(.)$	probability density function
$\kappa(.,.)$	Mercer kernel function
$K(.)$	kernel function for density estimation
$K_h(.)$	kernel function with width h
$G_h(.)$	Gaussian kernel function with width h
$\mathscr{H}_k$	reproducing kernel Hilbert space induced by Mercer kernel κ
$\mathbb{F}_\kappa$	feature space induced by Mercer kernel κ
W	weight vector
Ω	weight vector in feature space
$\tilde{W}$	weight error vector
η	step size
L	sliding data length
MSE	mean square error
LMS	least mean square
NLMS	normalized least mean square
LS	least squares
RLS	recursive least squares
MLE	maximum likelihood estimation
EM	expectation-maximization
FLOM	fractional lower order moment
LMP	least mean p-power
LAD	least absolute deviation
LMF	least mean fourth
FIR	finite impulse response
IIR	infinite impulse response
AR	auto regressive
ADALINE	adaptive linear neuron
MLP	multilayer perceptron
RKHS	reproducing kernel Hilbert space
KAF	kernel adaptive filtering
KLMS	kernel least mean square
KAPA	kernel affine projection algorithm
KMEE	kernel minimum error entropy
KMC	kernel maximum correntropy
PDF	probability density function
KDE	kernel density estimation
GGD	generalized Gaussian density
$S\alpha S$	symmetric α-stable
MEP	maximum entropy principle
DPI	data processing inequality
EPI	entropy power inequality
MEE	minimum error entropy
MCC	maximum correntropy criterion
IP	information potential
QIP	quadratic information potential
CRE	cumulative residual entropy
SIP	survival information potential
QSIP	survival quadratic information potential

KLID	Kullback—Leibler information divergence
EDC	Euclidean distance criterion
MinMI	minimum mutual information
MaxMI	maximum mutual information
AIC	Akaike's information criterion
BIC	Bayesian information criterion
MDL	minimum description length
FIM	Fisher information matrix
FIRM	Fisher information rate matrix
MIH	minimum identifiable horizon
ITL	information theoretic learning
BIG	batch information gradient
FRIG	forgetting recursive information gradient
SIG	stochastic information gradient
SIDG	stochastic information divergence gradient
SMIG	stochastic mutual information gradient
FP	fixed point
FP-MEE	fixed-point minimum error entropy
RFP-MEE	recursive fixed-point minimum error entropy
EDA	estimation of distribution algorithm
SNR	signal to noise ratio
WEP	weight error power
EMSE	excess mean square error
IEP	intrinsic error power
ICA	independent component analysis
BSS	blind source separation
CRLB	Cramer—Rao lower bound
AEC	acoustic echo canceller

1 Introduction

1.1 Elements of System Identification

Mathematical models of systems (either natural or man-made) play an essential role in modern science and technology. Roughly speaking, a mathematical model can be imagined as a mathematical law that links the system inputs (causes) with the outputs (effects). The applications of mathematical models range from simulation and prediction to control and diagnosis in heterogeneous fields. System identification is a widely used approach to build a mathematical model. It estimates the model based on the observed data (usually with uncertainty and noise) from the unknown system.

Many researchers try to provide an explicit definition for system identification. In 1962, Zadeh gave a definition as follows [1]: "System identification is the determination, on the basis of observations of input and output, of a system within a specified class of systems to which the system under test is equivalent." It is almost impossible to find out a model completely matching the physical plant. Actually, the system input and output always include certain noises; the identification model is therefore only an approximation of the practical plant. Eykhoff [2] pointed out that the system identification tries to use a model to describe the essential characteristic of an objective system (or a system under construction), and the model should be expressed in a useful form. Clearly, Eykhoff did not expect to obtain an exact mathematical description, but just to create a model suitable for applications. In 1978, Ljung [3] proposed another definition: "The identification procedure is based on three entities: the data, the set of models, and the criterion. Identification, then, is to select the model in the model set that describes the data best, according to the criterion."

According to the definitions by Zadeh and Ljung, system identification consists of three elements (see Figure 1.1): data, model, and equivalence criterion (equivalence is often defined in terms of a criterion or a loss function). The three elements directly govern the identification performance, including the identification accuracy, convergence rate, robustness, and computational complexity of the identification algorithm [4]. How to optimally design or choose these elements is very important in system identification.

The model selection is a crucial step in system identification. Over the past decades, a number of model structures have been suggested, ranging from the simple

System Parameter Identification. DOI: http://dx.doi.org/10.1016/B978-0-12-404574-3.00001-4
© 2013 Tsinghua University Press Ltd. Published by Elsevier Inc. All rights reserved.

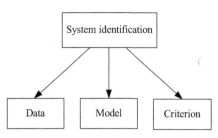

Figure 1.1 Three elements of system identification.

linear structures [FIR (finite impulse response), AR (autoregressive), ARMA (autoregressive and moving average), etc.] to more general nonlinear structures [NAR (nonlinear autoregressive), MLP (multilayer perceptron), RBF (radial basis function), etc.]. In general, model selection is a trade-off between the quality and the complexity of the model. In most practical situations, some prior knowledge may be available regarding the appropriate model structure or the designer may wish to limit to a particular model structure that is tractable and meanwhile can make a good approximation to the true system. Various model selection criteria have also been introduced, such as the cross-validation (CV) criterion [5], Akaike's information criterion (AIC) [6,7], Bayesian information criterion (BIC) [8], and minimum description length (MDL) criterion [9,10].

The data selection (the choice of the measured variables) and the optimal input design (experiment design) are important issues. The goal of experiment design is to adjust the experimental conditions so that maximal information is gained from the experiment (such that the measured data contain the maximal information about the unknown system). The optimality criterion for experiment design is usually based on the information matrices [11]. For many nonlinear models (e.g., the kernel-based model), the input selection can significantly help to reduce the network size [12].

The choice of the equivalence criterion (or approximation criterion) is another key issue in system identification. The approximation criterion measures the difference (or similarity) between the model and the actual system, and allows determination of how good the estimate of the system is. Different choices of the approximation criterion will lead to different estimates. The task of parametric system identification is to adjust the model parameters such that a predefined approximation criterion is minimized (or maximized). As a measure of accuracy, the approximation criterion determines the performance surface, and has significant influence on the optimal solutions and convergence behaviors. The development of new identification approximation criteria is an important emerging research topic and this will be the focus of this book.

It is worth noting that many machine learning methods also involve three elements: model, data, and optimization criterion. Actually, system identification can be viewed, to some extent, as a special case of supervised machine learning. The main terms in system identification and machine learning are reported in Table 1.1. In this book, these terminologies are used interchangeably.

Table 1.1 Main Terminologies in System Identification and Machine Learning

System Identification	Machine Learning
Model, filter	Learning machine, network
Parameters, coefficients	Weights
Identify, estimate	Learn, train
Observations, measurements	Examples, training data
Overparametrization	Overtraining, overfitting

1.2 Traditional Identification Criteria

Traditional identification (or estimation) criteria mainly include the least squares (LS) criterion [13], minimum mean square error (MMSE) criterion [14], and the maximum likelihood (ML) criterion [15,16]. The LS criterion, defined by minimizing the sum of squared errors (an error being the difference between an observed value and the fitted value provided by a model), could at least dates back to Carl Friedrich Gauss (1795). It corresponds to the ML criterion if the experimental errors have a Gaussian distribution. Due to its simplicity and efficiency, the LS criterion has been widely used in problems, such as estimation, regression, and system identification. The LS criterion is mathematically tractable, and the linear LS problem has a closed form solution. In some contexts, a regularized version of the LS solution may be preferable [17]. There are many identification algorithms developed with LS criterion. Typical examples are the recursive least squares (RLS) and its variants [4]. In statistics and signal processing, the MMSE criterion is a common measure of estimation quality. An MMSE estimator minimizes the mean square error (MSE) of the fitted values of a dependent variable. In system identification, the MMSE criterion is often used as a criterion for stochastic approximation methods, which are a family of iterative stochastic optimization algorithms that attempt to find the extrema of functions which cannot be computed directly, but only estimated via noisy observations. The well-known least mean square (LMS) algorithm [18−20], invented in 1960 by Bernard Widrow and Ted Hoff, is a stochastic gradient descent algorithm under MMSE criterion. The ML criterion is recommended, analyzed, and popularized by R.A. Fisher [15]. Given a set of data and underlying statistical model, the method of ML selects the model parameters that maximize the likelihood function (which measures the degree of "agreement" of the selected model with the observed data). The ML estimation provides a unified approach to estimation, which corresponds to many well-known estimation methods in statistics. The ML parameter estimation possesses a number of attractive limiting properties, such as consistency, asymptotic normality, and efficiency.

The above identification criteria (LS, MMSE, ML) perform well in most practical situations, and so far are still the workhorses of system identification. However, they have some limitations. For example, the LS and MMSE capture only the second-order statistics in the data, and may be a poor approximation criterion, especially in nonlinear and non-Gaussian (e.g., heavy tail or finite range distributions) situations. The ML criterion requires the knowledge of the conditional distribution (likelihood function) of the data given parameters, which is unavailable in many practical problems. In some complicated problems, the ML estimators are unsuitable or do not exist. Thus, selecting a new criterion beyond second-order statistics and likelihood function is attractive in problems of system identification.

In order to take into account higher order (or lower order) statistics and to select an optimal criterion for system identification, many researchers studied the non-MSE (nonquadratic) criteria. In an early work [21], Sherman first proposed the non-MSE criteria, and showed that in the case of Gaussian processes, a large family of non-MSE criteria yields the same predictor as the linear MMSE predictor of Wiener. Later, Sherman's results and several extensions were revisited by Brown [22], Zakai [23], Hall and Wise [24], and others. In [25], Ljung and Soderstrom discussed the possibility of a general error criterion for recursive parameter identification, and found an optimal criterion by minimizing the asymptotic covariance matrix of the parameter estimates. In [26,27], Walach and Widrow proposed a method to select an optimal identification criterion from the least mean fourth (LMF) family criteria. In their approach, the optimal choice is determined by minimizing a cost function which depends on the moments of the interfering noise. In [28], Douglas and Meng utilized the calculus of variations method to solve the optimal criterion among a large family of general error criteria. In [29], Al-Naffouri and Sayed optimized the error nonlinearity (derivative of the general error criterion) by optimizing the steady state performance. In [30], Pei and Tseng investigated the least mean p-power (LMP) criterion. The fractional lower order moments (FLOMs) of the error have also been used in adaptive identification in the presence of impulse alpha-stable noises [31,32]. Other non-MSE criteria include the M-estimation criterion [33], mixed norm criterion [34−36], risk-sensitive criterion [37,38], high-order cumulant (HOC) criterion [39−42], and so on.

1.3 Information Theoretic Criteria

Information theory is a branch of statistics and applied mathematics, which is exactly created to help studying the theoretical issues of optimally encoding messages according to their statistical structure, selecting transmission rates according to the noise levels in the channel, and evaluating the minimal distortion in messages [43]. Information theory was first developed by Claude E. Shannon to find fundamental limits on signal processing operations like compressing data and on reliably storing and communicating data [44]. After the pioneering work of Shannon, information theory found applications in many scientific areas, including physics, statistics, cryptography, biology, quantum computing, and so on. Moreover, information theoretic measures (entropy, divergence, mutual information, etc.) and principles (e.g., the principle of maximum entropy) were widely used in engineering areas, such as signal processing, machine learning, and other forms of data analysis. For example, the maximum entropy spectral analysis (MaxEnt spectral analysis) is a method of improving spectral estimation based on the principle of maximum entropy [45−48]. MaxEnt spectral analysis is based on choosing the spectrum which corresponds to the most random or the most unpredictable time series whose autocorrelation function agrees with the known values. This assumption, corresponding to the concept of maximum entropy as used in both statistical mechanics and information theory, is maximally noncommittal with respect to the unknown values of the autocorrelation function of the time series. Another example is the Infomax principle, an optimization principle for neural networks and other information processing systems, which prescribes that a function that maps a set of input values to a set of output values should be chosen or learned so as to maximize the average mutual information between input and output [49−53]. Information theoretic methods (such as Infomax) were successfully used in independent component analysis (ICA) [54−57] and blind source separation (BSS) [58−61]. In recent years, Jose C. Principe and his coworkers studied systematically the application of information theory to adaptive signal processing and machine learning [62−68]. They proposed the concept of *information theoretic learning* (ITL), which is achieved with information theoretic descriptors of entropy and dissimilarity (divergence and mutual information) combined with nonparametric density estimation. Their studies show that the ITL can bring robustness and generality to the cost function and improve the learning performance. One of the appealing features of ITL is that it can, with minor modifications, use the conventional learning algorithms of adaptive filters, neural networks, and kernel learning. The ITL links information theory, nonparametric estimators, and reproducing kernel Hilbert spaces (RKHS) in a simple and unconventional way [64]. A unifying framework of ITL is presented in Appendix A, such that the readers can easily understand it (for more details, see [64]).

Information theoretic methods have also been suggested by many authors for the solution of the related problems of system identification. In an early work [69], Zaborszky showed that information theory could provide a unifying viewpoint for the general identification problem. According to [69], the unknown parameters that need to be identified may represent the output of an information source which is transmitted over a channel, a specific identification technique. The identified values of the parameters are the output of the information channel represented by the identification technique. An identification technique can then be judged by its properties as an information channel transmitting the information contained in the parameters to be identified. In system parameter identification, the inverse of the Fisher information provides a lower bound (also known as the Cramér−Rao lower bound) on the variance of the estimator [70−74]. The rate distortion function in information theory can also be used to obtain the performance limitations in parameter estimation [75−79]. Many researchers also showed that there are elegant relationships between information theoretic measures (entropy, divergence, mutual information, etc.) and classical identification criteria like the MSE [80−85]. More importantly, many studies (especially those in ITL) suggest that information theoretic measures of entropy and divergence can be used as an identification criterion (referred to as the "information theoretic criterion," or simply, the "information criterion"), and can improve identification performance in many realistic scenarios. The choice of information theoretic criteria is very natural and reasonable since they capture higher order statistics and information content of signals rather than simply their energy. The information theoretic criteria and related identification algorithms are the main content of this book. Some of the content of this book had appeared in the ITL book (by Jose C. Principe) published in 2010 [64].

In this book, we mainly consider three kinds of information criteria: the minimum error entropy (MEE) criteria, the minimum information divergence criteria, and the mutual information-based criteria. Below, we give a brief overview of the three kinds of criteria.

1.3.1   MEE Criteria

Entropy is a central quantity in information theory, which quantifies the average uncertainty involved in predicting the value of a random variable. As the entropy measures the average uncertainty contained in a random variable, its minimization makes the distribution more concentrated. In [79,86], Weidemann and Stear studied the parameter estimation for nonlinear and non-Gaussian discrete-time systems by using the error entropy as the criterion functional, and proved that the reduced error entropy is upper bounded by the amount of information obtained by observation. Later, Tomita et al. [87] and Kalata and Priemer [88] applied the MEE criterion to study the optimal filtering and smoothing estimators, and provided a new interpretation for the filtering and smoothing problems from an information theoretic viewpoint. In [89], Minamide extended Weidemann and Stear's results to the continuous-time estimation models. The MEE estimation was reformulated by Janzura et al. as a problem of finding the optimal locations of probability densities in a given mixture such that the resulting entropy is minimized [90]. In [91], the minimum entropy of a mixture of conditional symmetric and unimodal (CSUM) distributions was studied. Some important properties of the MEE estimation were also reported in [92−95].

In system identification, when the errors (or residuals) are not Gaussian distributed, a more appropriate approach would be to constrain the error entropy [64]. The evaluation of the error entropy, however, requires the knowledge of the data distributions, which are usually unknown in practical applications. The nonparametric kernel (Parzen window) density estimation [96−98] provides an efficient way to estimate the error entropy directly from the error samples. This approach has been successfully applied in ITL and has the added advantages of linking information theory, adaptation, and kernel methods [64]. With kernel density estimation (KDE), Renyi's quadratic entropy can be easily calculated by a double sum over error samples [64]. The argument of the log in quadratic Renyi entropy estimator is named the *quadratic information potential* (QIP) estimator. The QIP is a central criterion function in ITL [99−106]. The computationally simple, nonparametric entropy estimators yield many well-behaved gradient algorithms to identify the system parameters such that the error entropy is minimized [64]. It is worth noting that the MEE criterion can also be used to identify the system structure. In [107], the Shannon's entropy power reduction ratio (EPRR) was introduced to select the terms in orthogonal forward regression (OFR) algorithms.

1.3.2 Minimum Information Divergence Criteria

An information divergence (say the Kullback—Leibler information divergence [108]) measures the dissimilarity between two distributions, which is useful in the analysis of parameter estimation and model identification techniques. A natural way of system identification is to minimize the information divergence between the actual (empirical) and model distributions of the data [109]. In an early work [7], Akaike suggested the use of the Kullback—Leibler divergence (KL-divergence) criterion via its sensitivity to parameter variations, showed its applicability to various statistical model fitting problems, and related it to the ML criterion. The AIC and its variants have been extensively studied and widely applied in problems of model selection [110—114]. In [115], Baram and Sandell employed a version of KL-divergence, which was shown to possess the property of being a metric on the parameter set, to treat the identification and modeling of a dynamical system, where the model set under consideration does not necessarily include the observed system. The minimum information divergence criterion has also been applied to study the simplification and reduction of a stochastic system model [116—119]. In [120], the problem of parameter identifiability with KL-divergence criterion was studied. In [121,122], several sequential (online) identification algorithms were developed to minimize the KL-divergence and deal with the case of incomplete data. In [123,124], Stoorvogel and Schuppen studied the identification of stationary Gaussian processes, and proved that the optimal solution to an approximation problem for Gaussian systems with the divergence criterion is identical to the main step of the subspace algorithm. In [125,126], motivated by the idea of *shaping the probability density function (PDF)*, the divergence between the actual error distribution and a reference (or target) distribution was used as an identification criterion. Some extensions of the KL-divergence, such as the α-divergence or ϕ-divergence, can also be employed as a criterion function for system parameter estimation [127—130].

1.3.3 Mutual Information-Based Criteria

Mutual information measures the statistical dependence between random variables. There are close relationships between mutual information and MMSE estimation. In [80], Duncan showed that for a continuous-time additive white Gaussian noise channel, the minimum mean square filtering (causal estimation) error is twice the input—output mutual information for any underlying signal distribution. Moreover, in [81], Guo et al. showed that the derivative of the mutual information was equal to half the MMSE in noncausal estimation. Like the entropy and information divergence, the mutual information can also be employed as an identification criterion. Weidemann and Stear [79], Janzura et al. [90], and Feng et al. [131] proved that minimizing the mutual information between estimation error and observations is equivalent to minimizing the error entropy. In [124], Stoorvogel and Schuppen showed that for a class of identification problems, the criterion of mutual information rate is identical to the criterion of exponential-of-quadratic cost and to H_∞ entropy (see [132] for the definition of H_∞ entropy). In [133], Yang and Sakai proposed a novel identification algorithm using ICA, which was derived by minimizing the mutual information between the estimated additive noise and the input signal. In [134], Durgaryan and Pashchenko proposed a consistent method of identification of systems by maximum mutual information (MaxMI) criterion and proved the conditions for identifiability. The MaxMI criterion has been successfully applied to identify the FIR and Wiener systems [135,136].

Besides the above-mentioned information criteria, there are many other information-based identification criteria, such as the maximum correntropy criterion (MCC) [137−139], minimization of error entropy with fiducial points (MEEF) [140], and minimum Fisher information criterion [141]. In addition to the AIC criterion, there are also many other information criteria for model selection, such as BIC [8] and MDL [9].

1.4   Organization of This Book

Up to now, considerable work has been done on system identification with information theoretic criteria, although the theory is still far from complete. So far there have been several books on the model selection with information critera (e.g., see [142−144]), but this book will provide a comprehensive treatment of system parameter identification with information criteria, with emphasis on the nonparametric cost functions and gradient-based identification algorithms. The rest of the book is organized as follows.

Chapter 2 presents the definitions and properties of some important information measures, including entropy, mutual information, information divergence, Fisher information, etc. This is a foundational chapter for the readers to understand the basic concepts that will be used in later chapters.

Chapter 3 reviews the information theoretic approaches for parameter estimation (classical and Bayesian), such as the maximum entropy estimation, minimum divergence estimation, and MEE estimation, and discusses the relationships between information theoretic methods and conventional alternatives. At the end of this chapter, a brief overview of several information criteria (AIC, BIC, MDL) for model selection is also presented. This chapter is vital for readers to understand the general theory of the information theoretic criteria.

Chapter 4 discusses extensively the system identification under MEE criteria. This chapter covers a brief sketch of system parameter identification, empirical error entropy criteria, several gradient-based identification algorithms, convergence analysis, optimization of the MEE criteria, survival information potential, and the Δ-entropy criterion. Many simulation examples are presented to illustrate the performance of the developed algorithms. This chapter ends with a brief discussion of system identification under the MCC.

Chapter 5 focuses on the system identification under information divergence criteria. The problem of parameter identifiability under mimimum KL-divergence criterion is analyzed. Then, motivated by the idea of *PDF shaping*, we introduce the minimum information divergence criterion with a reference PDF, and develop the corresponding identification algorithms. This chapter ends with an adaptive infinite impulsive response (IIR) filter with Euclidean distance criterion.

Chaper 6 changes the focus to the mutual information-based criteria: the mimimum mutual information (MinMI) criterion and the MaxMI criterion. The system identification under MinMI criterion can be converted to an ICA problem. In order to uniquely determine an optimal solution under MaxMI criterion, we propose a *double-criterion* identification method.

Appendix A: Unifying Framework of ITL

Figure A.1 shows a unifying framework of ITL (supervised or unsupervised). In Figure A.1, the cost $C(Y, D)$ denotes generally an information measure (entropy, divergence, or mutual information) between Y and D, where Y is the output of the model (learning machine) and D depends on which position the switch is in. ITL is then to adjust the parameters ω such that the cost $C(Y, D)$ is optimized (minimized or maximized).

1. Switch in position 1

   When the switch is in position 1, the cost involves the model output Y and an external desired signal Z. Then the learning is supervised, and the goal is to make the output signal and the desired signal as "close" as possible. In this case, the learning can be categorized into two categories: (a) filtering (or regression) and classification and (b) feature extraction.

   a. Filtering and classification

      In traditional filtering and classification, the cost function is in general the MSE or misclassification error rate (the $0-1$ loss). In ITL framework, the problem can be

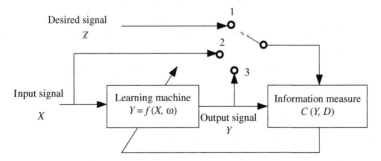

Figure A.1 Unifying ITL framework.

formulated as minimizing the divergence or maximizing the mutual information between output Y and the desired response Z, or minimizing the entropy of the error between the output and the desired responses (i.e., MEE criterion).

b. *Feature extraction*

In machine learning, when the input data are too large and the dimensionality is very high, it is necessary to transform nonlinearly the input data into a reduced representation set of features. Feature extraction (or feature selection) involves reducing the amount of resources required to describe a large set of data accurately. The feature set will extract the relevant information from the input in order to perform the desired task using the reduced representation instead of the full- size input. Suppose the desired signal is the class label, then an intuitive cost for feature extraction should be some measure of "relevance" between the projection outputs (features) and the labels. In ITL, this problem can be solved by maximizing the mutual information between the output Y and the label C.

2. Switch in position 2

When the switch is in position 2, the learning is in essence unsupervised because there is no external signal besides the input and output signals. In this situation, the well-known optimization principle is the *Maximum Information Transfer*, which aims to maximize the mutual information between the original input data and the output of the system. This principle is also known as the principle of maximum information preservation (Infomax). Another information optimization principle for unsupervised learning (clustering, principal curves, vector quantization, etc.) is the *Principle of Relevant Information* (PRI) [64]. The basic idea of PRI is to minimize the data redundancy (entropy) while preserving the similarity to the original data (divergence).

3. Switch in position 3

When the switch is in position 3, the only source of data is the model output, which in this case is in general assumed multidimensional. Typical examples of this case include ICA, clustering, output entropy optimization, and so on.

Independent component analysis: ICA is an unsupervised technique aiming to reduce the redundancy between components of the system output. Given a nonlinear multiple-input−multiple-output (MIMO) system $y = f(x, \omega)$, the nonlinear ICA usually optimizes the parameter vector ω such that the mutual information between the components of y is minimized.

Clustering: Clustering (or clustering analysis) is a common technique for statistical data analysis used in machine learning, pattern recognition, bioinformatics, etc. The goal of clustering is to divide the input data into groups (called clusters) so that the objects in the same cluster are more "similar" to each other than to those in other clusters, and different clusters are defined as compactly and distinctly as possible. Information theoretic measures, such as entropy and divergence, are frequently used as an optimization criterion for clustering.

Output entropy optimization: If the switch is in position 3, one can also optimize (minimize or maximize) the entropy at system output (usually subject to some constraint on the weight norm or nonlinear topology) so as to capture the underlying structure in high dimensional data.

4. Switch simultaneously in positions 1 and 2

In Figure A.1, the switch can be simultaneously in positions 1 and 2. In this case, the cost has access to input data X, output data Y, and the desired or reference data Z. A well-known example is the *Information Bottleneck* (IB) method, introduced by Tishby et al. [145]. Given a random variable X and an observed relevant variable Z, and assuming that the joint distribution between X and Z is known, the IB method aims to compress X and try to find the best trade-off between (i) the minimization of mutual information between X and its compressed version Y and (ii) the maximization of mutual information between Y and the relevant variable Z. The basic idea in IB is to find a reduced representation of X while preserving the information of X with respect to another variable Z.

2 Information Measures

The concept of information is so rich that there exist various definitions of information measures. Kolmogorov had proposed three methods for defining an information measure: probabilistic method, combinatorial method, and computational method [146]. Accordingly, information measures can be categorized into three categories: probabilistic information (or statistical information), combinatory information, and algorithmic information. This book focuses mainly on statistical information, which was first conceptualized by Shannon [44]. As a branch of mathematical statistics, the establishment of Shannon information theory lays down a mathematical framework for designing optimal communication systems. The core issues in Shannon information theory are how to measure the amount of information and how to describe the information transmission. According to the feature of data transmission in communication, Shannon proposed the use of entropy, which measures the uncertainty contained in a probability distribution, as the definition of information in the data source.

2.1 Entropy

Definition 2.1 Given a discrete random variable X with probability mass function $P\{X = x_k\} = p_k$, $k = 1, \ldots, n$, Shannon's (discrete) entropy is defined by [43]

$$H(X) = \sum_{k=1}^{n} p_k I(p_k) \tag{2.1}$$

where $I(p_k) = -\log p_k$ is Hartley's amount of information associated with the discrete value x_k with probability p_k.[1] This information measure was originally devised by Claude Shannon in 1948 to study the amount of information in a transmitted message. Shannon entropy measures the average information (or uncertainty) contained in a probability distribution and can also be used to measure many other concepts, such as diversity, similarity, disorder, and randomness. However, as the discrete entropy depends only on the distribution P, and takes no account of the values, it is independent of the dynamic range of the random variable. The discrete entropy is unable to differentiate between two random variables that have the same distribution but different dynamic ranges. Actually the discrete

---

[1] In this book, "log" always denotes the natural logarithm. The entropy will then be measured in nats.

System Parameter Identification. DOI: http://dx.doi.org/10.1016/B978-0-12-404574-3.00002-6
© 2013 Tsinghua University Press Ltd. Published by Elsevier Inc. All rights reserved.

random variables with the same entropy may have arbitrarily small or large variance, a typical measure for value dispersion of a random variable.

Since system parameter identification deals, in general, with continuous random variables, we are more interested in the entropy of a continuous random variable.

Definition 2.2 If X is a continuous random variable with PDF $p(x)$, $x \in C$, Shannon's differential entropy is defined as

$$H(X) = -\int_C p(x)\log p(x)\mathrm{d}x \tag{2.2}$$

The differential entropy is a functional of the PDF $p(x)$. For this reason, we also denote it by $H(p)$. The entropy definition in (2.2) can be extended to multiple random variables. The joint entropy of two continuous random variables X and Y is

$$H(X, Y) = -\int\int p(x, y)\log p(x, y)\mathrm{d}x\,\mathrm{d}y \tag{2.3}$$

where $p(x, y)$ denotes the joint PDF of (X, Y). Furthermore, one can define the conditional entropy of X given Y as

$$H(X|Y) = -\int\int p(x, y)\log p(x|y)\mathrm{d}x\,\mathrm{d}y \tag{2.4}$$

where $p(x|y)$ is the conditional PDF of X given Y.[2]

If X and Y are discrete random variables, the entropy definitions in (2.3) and (2.4) only need to replace the PDFs with the probability mass functions and the integral operation with the summation.

Theorem 2.1 Properties of the differential entropy[3] :

1. Differential entropy can be either positive or negative.
2. Differential entropy is not related to the mean value (shift invariant), i.e., $H(X + c) = H(X)$, where $c \in \mathbb{R}$ is an arbitrary constant.
3. $H(X, Y) = H(X) + H(Y|X) = H(Y) + H(X|Y)$.
4. $H(X|Y) \leq H(X)$, $H(Y|X) \leq H(Y)$.
5. Entropy has the concavity property: $H(p)$ is a concave function of p, that is, $\forall 0 \leq \lambda \leq 1$, we have

$$H(\lambda p_1 + (1 - \lambda)p_2) \geq \lambda H(p_1) + (1 - \lambda)H(p_2) \tag{2.5}$$

---

[2] Strictly speaking, we should use some subscripts to distinguish the PDFs $p(x)$, $p(x, y)$, and $p(x|y)$. For example, we can write them as $p_X(x)$, $p_{XY}(x, y)$, $p_{X|Y}(x|y)$. In this book, for simplicity we often omit these subscripts if no confusion arises.

[3] The detailed proofs of these properties can be found in related information theory textbooks, such as "Elements of Information Theory" written by Cover and Thomas [43].

6. If random variables X and Y are mutually independent, then

$$H(X + Y) \geq \max\{H(X), H(Y)\} \tag{2.6}$$

that is, the entropy of the sum of two independent random variables is no smaller than the entropy of each individual variable.

7. *Entropy power inequality (EPI)*: If X and Y are mutually independent d-dimensional random variables, we have

$$\exp\left\{\frac{2}{d}H(X+Y)\right\} \geq \exp\left\{\frac{2}{d}H(X)\right\} + \exp\left\{\frac{2}{d}H(Y)\right\} \tag{2.7}$$

with equality if and only if X and Y are Gaussian distributed and their covariance matrices are in proportion to each other.

8. Assume X and Y are two d-dimensional random variables, $Y = \psi(X)$, ψ denotes a smooth bijective mapping defined over $\mathbb{R}^d$, J_ψ is the Jacobi matrix of ψ, then

$$H(Y) = H(X) + \int_{\mathbb{R}^d} p(x)\log|\det J_\psi| dx \tag{2.8}$$

where det denotes the determinant.

9. Suppose X is a d-dimensional Gaussian random variable, $X \sim \mathcal{N}(\mu, \Sigma)$, i.e.,

$$p(x) = \frac{1}{(2\pi)^{d/2}\sqrt{\det\Sigma}}\exp\left\{-\frac{1}{2}(x-\mu)^T\Sigma^{-1}(x-\mu)\right\}, \quad x \in \mathbb{R}^d \tag{2.9}$$

Then the differential entropy of X is

$$H(X) = \frac{d}{2} + \frac{1}{2}\log\{(2\pi)^d\det\Sigma\} \tag{2.10}$$

Differential entropy measures the uncertainty and dispersion in a probability distribution. Intuitively, the larger the value of entropy, the more scattered the probability density of a random variable or in other word, the smaller the value of entropy, the more concentrated the probability density. For a one-dimensional random variable, the differential entropy is similar to the variance. For instance, the differential entropy of a one-dimensional Gaussian random variable X is $H(X) = (1/2) + (1/2)\log(2\pi \text{Var}(X))$, where $\text{Var}(X)$ denotes the variance of X. It is clear to see that in this case the differential entropy increases monotonically with increasing variance. However, the entropy is in essence quite different from the variance; it is a more comprehensive measure. The variance of some random variable is infinite, while the entropy is still finite. For example, consider the following Cauchy distribution[4] :

$$p(x) = \frac{\lambda}{\pi}\frac{1}{\lambda^2 + x^2}, \quad -\infty < x < \infty, \quad \lambda > 0 \tag{2.11}$$

---

[4] Cauchy distribution is a non-Gaussian α-stable distribution (see Appendix B).

Its variance is infinite, while the differential entropy is $\log(4\pi\lambda)$ [147].

There is an important entropy optimization principle, that is, the maximum entropy (MaxEnt) principle enunciated by Jaynes [148] and Kapur and Kesavan [149]. According to MaxEnt, among all the distributions that satisfy certain constraints, one should choose the distribution that maximizes the entropy, which is considered to be the most objective and most impartial choice. MaxEnt is a powerful and widely accepted principle for statistical inference with incomplete knowledge of the probability distribution.

The maximum entropy distribution under characteristic moment constraints can be obtained by solving the following optimization problem:

$$
\begin{cases}
\max_p H(p) = -\int_{\mathbb{R}} p(x) \log p(x) dx \\
\text{s.t.} \begin{cases} \int_{\mathbb{R}} p(x) dx = 1 \\ \int_{\mathbb{R}} g_k(x) p(x) dx = \mu_k, \quad k = 1, 2, \ldots, K \end{cases}
\end{cases}
\tag{2.12}
$$

where $\int_{\mathbb{R}} p(x) dx = 1$ is the natural constraint (the normalization constraint) and $\int_{\mathbb{R}} g_k(x) p(x) dx = \mu_k$ $(k = 1, 2, \ldots, K)$ denote K (generalized) characteristic moment constraints.

Theorem 2.2 (Maximum Entropy PDF) Satisfying the constraints in (2.12), the maximum entropy PDF is given by

$$
p_{\text{MaxEnt}}(x) = \exp\left(-\lambda_0 - \sum_{k=1}^{K} \lambda_k g_k(x)\right)
\tag{2.13}
$$

where the coefficients $\lambda_i (i = 0, 1, \ldots, K)$ are the solution of the following equations[5]:

$$
\begin{cases}
\int_{\mathbb{R}} \exp\left(-\sum_{k=1}^{K} \lambda_k g_k(x)\right) dx = \exp(\lambda_0) \\
\dfrac{\int_{\mathbb{R}} g_i(x) \exp\left(-\sum_{k=1}^{K} \lambda_k g_k(x)\right) dx}{\exp(\lambda_0)} = \mu_i, \quad i = 1, 2, \ldots, K
\end{cases}
\tag{2.14}
$$

In statistical information theory, in addition to Shannon entropy, there are many other definitions of entropy, such as Renyi entropy (named after Alfred Renyi) [152], Havrda−Charvat entropy [153], Varma entropy [154], Arimoto entropy [155], and (h, ϕ)-entropy [156]. Among them, (h, ϕ)-entropy is the most generalized definition of entropy. (h, ϕ)-entropy of a continuous random variable X is defined by [156]

$$
H_\phi^h(X) = h\left(\int_{-\infty}^{+\infty} \phi[p(x)] dx\right)
\tag{2.15}
$$

---

[5] On how to solve these equations, interested readers are referred to [150,151].

where either $\phi:[0, \infty) \to \mathbb{R}$ is a concave function and $h:\mathbb{R} \to \mathbb{R}$ is a monotonously increasing function or $\phi:[0, \infty) \to \mathbb{R}$ is a convex function and $h:\mathbb{R} \to \mathbb{R}$ is a monotonously decreasing function. When $h(x) = x$, (h, ϕ)-entropy becomes the ϕ-entropy:

$$H_\phi(X) = \int_{-\infty}^{+\infty} \phi[p(x)]dx \tag{2.16}$$

where $\phi:[0, \infty) \to \mathbb{R}$ is a concave function. Similar to Shannon entropy, (h, ϕ)-entropy is also shift-invariant and satisfies (see Appendix C for the proof)

$$H_\phi^h(X + Y) \geq \max\{H_\phi^h(X), H_\phi^h(Y)\} \tag{2.17}$$

where X and Y are two mutually independent random variables. Some typical examples of (h, ϕ)-entropy are given in Table 2.1. As one can see, many entropy definitions can be regarded as the special cases of (h, ϕ)-entropy.

From Table 2.1, Renyi's entropy of order-α is defined as

$$\begin{aligned} H_\alpha(X) &= \frac{1}{1 - \alpha} \log \int_{-\infty}^{\infty} p^\alpha(x)dx \\ &= \frac{1}{1 - \alpha} \log V_\alpha(X) \end{aligned} \tag{2.18}$$

where $\alpha > 0$, $\alpha \neq 1$, $V_\alpha(X) \triangleq \int_{-\infty}^{\infty} p^\alpha(x)dx$ is called the order-α information potential (when $\alpha = 2$, called the quadratic information potential, QIP) [64]. The Renyi entropy is a generalization of Shannon entropy. In the limit $\alpha \to 1$, it will converge to Shannon entropy, i.e., $\lim_{\alpha \to 1} H_\alpha(X) = H(X)$.

Table 2.1 (h, ϕ)-Entropies with Different h and ϕ Functions [130]

$h(x)$	$\phi(x)$	(h,ϕ)-entropy
x	$-x \log x$	Shannon (1948)
$(1-\alpha)^{-1} \log x$	x^α	Renyi (1961) ($\alpha > 0$, $\alpha \neq 1$)
$(m(m-r))^{-1} \log x$	$x^{r/m}$	Varma (1966) ($0 < r < m$, $m \geq 1$)
x	$(1-s)^{-1}(x^s - x)$	Havrda−Charvat (1967) ($s \neq 1$, $s > 0$)
$(t-1)^{-1}(x^t - 1)$	$x^{1/t}$	Arimoto (1971) ($t > 0, t \neq 1$)
x	$(1-s)^{-1}(x^s + (1-x)^s - 1)$	Kapur (1972) ($s \neq 1$)
$(1-s)^{-1}[\exp((s - 1)x) - 1]$	$x \log x$	Sharma and Mittal (1975) ($s > 0, s \neq 1$)
$(1 + (1/\lambda))\log(1 + \lambda) - (x/\lambda)$	$(1 + \lambda x)\log(1 + \lambda x)$	Ferreri (1980) ($\lambda > 0$)

   The previous entropies are all defined based on the PDFs (for continuous random variable case). Recently, some researchers also propose to define the entropy measure using the distribution or survival functions [157,158]. For example, the cumulative residual entropy (CRE) of a scalar random variable X is defined by [157]

$$\varepsilon(X) = -\int_{R_+} \overline{F}_{|X|}(x) \log \overline{F}_{|X|}(x) \mathrm{d}x \tag{2.19}$$

where $\overline{F}_{|X|}(x) = P(|X| > x)$ is the survival function of $|X|$. The CRE is just defined by replacing the PDF with the survival function (of an absolute value transformation of X) in the original differential entropy (2.2). Further, the order-α ($\alpha > 0$) survival information potential (SIP) is defined as [159]

$$S_\alpha(X) = \int_{R_+} \overline{F}_{|X|}^\alpha(x) \mathrm{d}x \tag{2.20}$$

   This new definition of information potential is valid for both discrete and continuous random variables.
   In recent years, the concept of *correntropy* has also been applied successfully in signal processing and machine learning [137]. The correntropy is not a true entropy measure, but in this book it is still regarded as an information theoretic measure since it is closely related to Renyi's quadratic entropy (H_2), that is, the negative logarithm of the sample mean of correntropy (with Gaussian kernel) yields the Parzen estimate of Renyi's quadratic entropy [64]. Let X and Y be two random variables with the same dimensions, the correntropy is defined by

$$V(X, Y) = E[\kappa(X, Y)] = \int \kappa(x, y) \mathrm{d}F_{XY}(x, y) \tag{2.21}$$

where E denotes the expectation operator, $\kappa(.,.)$ is a translation invariant Mercer kernel[6], and $F_{XY}(x, y)$ denotes the joint distribution function of (X, Y). According to Mercer's theorem, any Mercer kernel $\kappa(.,.)$ induces a nonlinear mapping $\varphi(.)$ from the input space (original domain) to a high (possibly infinite) dimensional feature space F (a vector space in which the input data are embedded), and the inner product of two points $\varphi(X)$ and $\varphi(Y)$ in F can be implicitly computed by using the

---

[6] Let $(\mathscr{X}, \Sigma)$ be a measurable space and assume a real-valued function $\kappa(.,.)$ is defined on $\mathscr{X} \times \mathscr{X}$, i.e., $\kappa: \mathscr{X} \times \mathscr{X} \to R$. Then function $\kappa(.,.)$ is called a *Mercer kernel* if and only if it is a continuous, symmetric, and *positive-definite* function. Here, κ is said to be positive-definite if and only if

$$\iint \kappa(x, y) \mathrm{d}\mu(x) \mathrm{d}\mu(y) \geq 0$$

where μ denotes any finite signed Borel measure, $\mu: \Sigma \to R$. If the equality holds only for zero measure, then κ is said to be *strictly positive-definite* (SPD).

Mercer kernel (the so-called "kernel trick") [160−162]. Then the correntropy (2.21) can alternatively be expressed as

$$V(X, Y) = E[\langle \varphi(X), \varphi(Y)\rangle_F] \tag{2.22}$$

where $\langle ., .\rangle_F$ denotes the inner product in F. From (2.22), one can see that the correntropy is in essence a new measure of the similarity between two random variables, which generalizes the conventional correlation function to feature spaces.

2.2 Mutual Information

Definition 2.3 The mutual information between continuous random variables X and Y is defined as

$$I(X; Y) = \int \int p(x, y) \log \frac{p(x, y)}{p(x)p(y)} \, dx \, dy = E\left(\log \frac{p(X, Y)}{p(X)p(Y)} \right) \tag{2.23}$$

The conditional mutual information between X and Y, conditioned on random variable Z, is given by

$$I(X; Y|Z) = \int \int \int p(x, y, z) \log \frac{p(x, y|z)}{p(x|z)p(y|z)} \, dx \, dy \, dz \tag{2.24}$$

For a random vector[7] $X = [X_1, X_2, \ldots, X_n]^T$ $(n \geq 2)$, the mutual information between components is

$$I(X) = \sum_{i=1}^{n} H(X_i) - H(X) \tag{2.25}$$

Theorem 2.3 Properties of the mutual information:

1. Symmetry, i.e., $I(X; Y) = I(Y; X)$.
2. Non-negative, i.e., $I(X; Y) \geq 0$, with equality if and only if X and Y are mutually independent.
3. *Data processing inequality* (DPI): If random variables X, Y, Z form a Markov chain $X \rightarrow Y \rightarrow Z$, then $I(X; Y) \geq I(X; Z)$. Especially, if Z is a function of Y, $Z = \beta(Y)$, where $\beta(.)$ is a measurable mapping from Y to Z, then $I(X; Y) \geq I(X; \beta(Y))$, with equality if β is invertible and β^{-1} is also a measurable mapping.

---

[7] Unless mentioned otherwise, in this book a vector refers to a column vector.

4. The relationship between mutual information and entropy:

$$I(X;Y) = H(X) - H(X|Y), I(X;Y|Z) = H(X|Z) - H(X|YZ) \tag{2.26}$$

5. *Chain rule*: Let $Y_1, Y_2, \ldots, Y_l$ be l random variables. Then

$$I(X;Y_1, \ldots, Y_l) = I(X;Y_1) + \sum_{i=2}^{l} I(X;Y_i|Y_1, \ldots, Y_{i-1}) \tag{2.27}$$

6. If X, Y, (X, Y) are k, l, and $k + l$-dimension Gaussian random variables with, respectively, covariance matrices A, B, and C, then the mutual information between X and Y is

$$I(X;Y) = \frac{1}{2}\log\frac{\det A \det B}{\det C} \tag{2.28}$$

In particular, if $k = l = 1$, we have

$$I(X;Y) = -\frac{1}{2}\log(1 - \rho^2(X,Y)) \tag{2.29}$$

where $\rho(X, Y)$ denotes the correlation coefficient between X and Y.
7. Relationship between mutual information and MSE: Assume X and Y are two Gaussian random variables, satisfying $Y = \sqrt{\text{snr}}X + N$, where $\text{snr} \geq 0$, $N \sim \mathcal{N}(0, 1)$, N and X are mutually independent. Then we have [81]

$$\frac{d}{d\text{snr}}I(X;Y) = \frac{1}{2}\text{mmse}(X|Y) \tag{2.30}$$

where $\text{mmse}(X|Y)$ denotes the minimum MSE when estimating X based on Y.

Mutual information is a measure of the amount of information that one random variable contains about another random variable. The stronger the dependence between two random variables, the greater the mutual information is. If two random variables are mutually independent, the mutual information between them achieves the minimum zero. The mutual information has close relationship with the correlation coefficient. According to (2.29), for two Gaussian random variables, the mutual information is a monotonically increasing function of the correlation coefficient. However, the mutual information and the correlation coefficient are different in nature. The mutual information being zero implies that the random variables are mutually independent, thereby the correlation coefficient is also zero, while the correlation coefficient being zero does not mean the mutual information is zero (i.e., the mutual independence). In fact, the condition of independence is much stronger than mere uncorrelation. Consider the following Pareto distributions [149]:

$$\begin{cases} p_X(x) = a\theta_1^a x^{-(a+1)} \\ p_Y(y) = a\theta_2^a y^{-(a+1)} \\ p_{XY}(x,y) = -\frac{a(a+1)}{\theta_1\theta_2}\left(\frac{x}{\theta_1} + \frac{y}{\theta_2} - 1\right)^{-(a+2)} \end{cases} \tag{2.31}$$

where $\alpha > 1$, $x \geq \theta_1$, $y \geq \theta_2$. One can calculate $E[X] = a\theta_1/(a-1)$, $E[Y] = a\theta_2/(a-1)$, and $E[XY] = a^2\theta_1\theta_2/(a-1)^2$, and hence $\rho(X,Y) = 0$ (X and Y are uncorrelated). In this case, however, $p_{XY}(x,y) \neq p_X(x)p_Y(y)$, that is, X and Y are not mutually independent (the mutual information not being zero).

With mutual information, one can define the rate distortion function and the distortion rate function. The rate distortion function $R(D)$ of a random variable X with MSE distortion is defined by

$$R(D) = \inf_Y \{I(X;Y) : E[(X-Y)^2] \leq D^2\} \tag{2.32}$$

At the same time, the distortion rate function is defined as

$$D(R) = \inf_Y \left\{ \sqrt{E[(X-Y)^2]} : I(X;Y) \leq R \right\} \tag{2.33}$$

Theorem 2.4 If X is a Gaussian random variable, $X \sim \mathcal{N}(\mu, \sigma^2)$, then

$$\begin{cases} R(D) = \dfrac{1}{2}\log\left\{ \max\left(1, \dfrac{\sigma^2}{D^2}\right) \right\}, & D \geq 0 \\[2ex] D(R) = \sigma\exp(-R), & R \geq 0 \end{cases} \tag{2.34}$$

2.3 Information Divergence

In statistics and information geometry, an information divergence measures the "distance" of one probability distribution to the other. However, the divergence is a much weaker notion than that of the distance in mathematics, in particular it need not be symmetric and need not satisfy the triangle inequality.

Definition 2.4 Assume that X and Y are two random variables with PDFs $p(x)$ and $q(y)$ with common support. The Kullback–Leibler information divergence (KLID) between X and Y is defined by

$$D_{KL}(X\|Y) = D_{KL}(p\|q) = \int p(x)\log\frac{p(x)}{q(x)}dx \tag{2.35}$$

In the literature, the KL-divergence is also referred to as the discrimination information, the cross entropy, the relative entropy, or the directed divergence.

Theorem 2.5 Properties of KL-divergence:

1. $D_{KL}(p\|q) \geq 0$, with equality if and only if $p(x) = q(x)$.
2. Nonsymmetry: In general, we have $D_{KL}(p\|q) \neq D_{KL}(q\|p)$.
3. $D_{KL}(p(x,y)\|p(x)p(y)) = I(X;Y)$, that is, the mutual information between two random variables is actually the KL-divergence between the joint probability density and the product of the marginal probability densities.
4. Convexity property: $D_{KL}(p\|q)$ is a convex function of (p,q), i.e., $\forall\, 0 \leq \lambda \leq 1$, we have

$$D_{KL}(\overline{p}\|\overline{q}) \leq \lambda D_{KL}(p_1\|q_1) + (1-\lambda)D_{KL}(p_2\|q_2) \tag{2.36}$$

   where $\overline{p} = \lambda p_1 + (1-\lambda)p_2$ and $\overline{q} = \lambda q_1 + (1-\lambda)q_2$.
5. *Pinsker's inequality*: Pinsker inequality is an inequality that relates KL-divergence and the total variation distance. It states that

$$D_{KL}(p\|q) \geq \frac{1}{2}\left(\int |p(x) - q(x)|\mathrm{d}x\right)^2 \tag{2.37}$$

6. *Invariance under invertible transformation*: Given random variables X and Y, and the invertible transformation T, the KL-divergence remains unchanged after the transformation, i.e., $D_{KL}(X\|Y) = D_{KL}(T(X)\|T(Y))$. In particular, if $T(X) = X + c$, where c is a constant, then the KL-divergence is shift-invariant:

$$D_{KL}(X\|Y) = D_{KL}(X + c\|Y + c) \tag{2.38}$$

7. If X and Y are two d-dimensional Gaussian random variables, $X \sim \mathcal{N}(\mu_1, \Sigma_1)$, $Y \sim \mathcal{N}(\mu_2, \Sigma_2)$, then

$$D_{KL}(X\|Y) = \frac{1}{2}\left\{\log\frac{\det \Sigma_2}{\det \Sigma_1} + Tr(\Sigma_1(\Sigma_2^{-1} - \Sigma_1^{-1})) + (\mu_1 - \mu_2)^T\Sigma_2^{-1}(\mu_1 - \mu_2)\right\} \tag{2.39}$$

where Tr denotes the trace operator.

There are many other definitions of information divergence. Some quadratic divergences are frequently used in machine learning, since they involve only a simple quadratic form of PDFs. Among them, the Euclidean distance (ED) in probability spaces and the Cauchy–Schwarz (CS)-divergence are popular, and are defined respectively as [64]

$$D_{ED}(p\|q) = \int (p(x) - q(x))^2 \mathrm{d}x \tag{2.40}$$

$$D_{CS}(p\|q) = -\log\frac{\left(\int p(x)q(x)\mathrm{d}x\right)^2}{\int p^2(x)\mathrm{d}x \int q^2(x)\mathrm{d}x} \tag{2.41}$$

Clearly, the ED in (2.40) can be expressed in terms of QIP:

$$D_{ED}(p\|q) = V_2(p) + V_2(q) - 2V_2(p;q) \tag{2.42}$$

where $V_2(p;q) \triangleq \int p(x)q(x)\mathrm{d}x$ is named the cross information potential (CIP). Further, the CS-divergence of (2.41) can also be rewritten in terms of Renyi's quadratic entropy:

$$D_{CS}(p\|q) = 2H_2(p;q) - H_2(p) - H_2(q) \tag{2.43}$$

where $H_2(p;q) = -\log \int p(x)q(x)\mathrm{d}x$ is called Renyi's quadratic cross entropy.

Also, there is a much generalized definition of divergence, i.e., the ϕ-divergence, which is defined as [130]

$$D_\phi(p\|q) = \int q(x)\phi\left(\frac{p(x)}{q(x)}\right)\mathrm{d}x, \quad \phi \in \Phi^* \tag{2.44}$$

where Φ^* is a collection of convex functions, $\forall \phi \in \Phi^*$, $\phi(1) = 0$, $0\phi(0/0) = 0$, and $0\phi(p/0) = \lim_{u \to \infty} \phi(u)/u$. When $\phi(x) = x \log x$ (or $\phi(x) = x \log x - x + 1$), the ϕ-divergence becomes the KL-divergence. It is easy to verify that the ϕ-divergence satisfies the properties (1), (4), and (6) in Theorem 2.5. Table 2.2 gives some typical examples of ϕ-divergence.

2.4 Fisher Information

The most celebrated information measure in statistics is perhaps the one developed by R.A. Fisher (1921) for the purpose of quantifying information in a distribution about the parameter.

Definition 2.5 Given a parameterized PDF $p_Y(y, \theta)$, where $y \in \mathbb{R}^N$, $\theta = [\theta_1, \theta_2, \ldots, \theta_d]^T$ is a d-dimensional parameter vector, and assuming $p_Y(y, \theta)$ is continuously differentiable with respect to θ, then the Fisher information matrix (FIM) with respect to θ is

$$J_F(\theta) = \int_{\mathbb{R}^N} \frac{1}{p_Y(y, \theta)} \left[\frac{\partial}{\partial \theta}p_Y(y, \theta)\right]\left[\frac{\partial}{\partial \theta}p_Y(y, \theta)\right]^T \mathrm{d}y \tag{2.45}$$

Table 2.2 ϕ-Divergences with Different ϕ-Functions [130]

$\phi(x)$	ϕ-Divergence
$x \log x - x + 1$	Kullback−Leibler (1959)
$(x-1)\log x$	J-Divergence
$(x-1)^2/2$	Pearson (1900), Kagan (1963)
$(x^{\lambda+1} - x - \lambda(x-1))/(\lambda(\lambda+1)), \lambda \neq 0, -1$	Power-divergence (1984)
$(x-1)^2/(x+1)^2$	Balakrishnan and Sanghvi (1968)
$|1 - x^\alpha|^{1/\alpha}, 0 < \alpha < 1$	Matusita (1964)
$|1 - x|^\alpha, \alpha \geq 1$	χ-Divergence (1973)

Clearly, the FIM $J_F(\theta)$, also referred to as the Fisher information, is a $d \times d$ matrix. If θ is a location parameter, i.e., $p_Y(y, \theta) = p(y - \theta)$, Fisher information will be

$$J_F(Y) = \int_{R^N} \frac{1}{p(y)} \left[\frac{\partial}{\partial y} p(y) \right] \left[\frac{\partial}{\partial y} p(y) \right]^T dy \tag{2.46}$$

The Fisher information of (2.45) can alternatively be written as

$$J_F(\theta) = \int_{R^N} p_Y(y, \theta) \left[\frac{\partial}{\partial \theta} \log p_Y(y, \theta) \right] \left[\frac{\partial}{\partial \theta} \log p_Y(y, \theta) \right]^T dy$$

$$= E_\theta \left\{ \left[\frac{\partial}{\partial \theta} \log p_Y(Y, \theta) \right] \left[\frac{\partial}{\partial \theta} \log p_Y(Y, \theta) \right]^T \right\} \tag{2.47}$$

where E_θ stands for the expectation with respect to $p_Y(y, \theta)$. From (2.47), one can see that the Fisher information measures the "average sensitivity" of the logarithm of PDF to the parameter θ or the "average influence" of the parameter θ on the logarithm of PDF. The Fisher information is also a measure of the minimum error in estimating the parameter of a distribution. This is illustrated in the following theorem.

Theorem 2.6 (Cramer−Rao Inequality) Let $p_Y(y, \theta)$ be a parameterized PDF, where $y \in R^N$, $\theta = [\theta_1, \theta_2, \ldots, \theta_d]^T$ is a d-dimensional parameter vector, and assume that $p_Y(y, \theta)$ is continuously differentiable with respect to θ. Denote $\hat{\theta}(Y)$ an unbiased estimator of θ based on Y, satisfying $E_{\theta_0}[\hat{\theta}(Y)] = \theta_0$, where θ_0 denotes the true value of θ. Then

$$P \triangleq E_{\theta_0}[(\hat{\theta}(Y) - \theta_0)(\hat{\theta}(Y) - \theta_0)^T] \geq J_F^{-1}(\theta_0) \tag{2.48}$$

where P is the covariance matrix of $\hat{\theta}(Y)$.

Cramer−Rao inequality shows that the inverse of the FIM provides a lower bound on the error covariance matrix of the parameter estimator, which plays a significant role in parameter estimation. A proof of the Theorem 2.6 is given in Appendix D.

2.5 Information Rate

The previous information measures, such as entropy, mutual information, and KL-divergence, are all defined for random variables. These definitions can be further extended to various information rates, which are defined for random processes.

Definition 2.6 Let $\{X_t \in \mathbb{R}^{m_1}, t \in Z\}$ and $\{Y_t \in \mathbb{R}^{m_2}, t \in Z\}$ be two discrete-time stochastic processes, and denote $X^n = [X_1^T, X_2^T, \ldots, X_n^T]^T$, $Y^n = [Y_1^T, Y_2^T, \ldots, Y_n^T]^T$. The entropy rate of the stochastic process $\{X_t\}$ is defined as

$$\overline{H}(\{X_t\}) = \lim_{n \to \infty} \frac{1}{n} H(X^n) \tag{2.49}$$

The mutual information rate between $\{X_t\}$ and $\{Y_t\}$ is defined by

$$\overline{I}(\{X_t\}; \{Y_t\}) = \lim_{n \to \infty} \frac{1}{n} I(X^n; Y^n) \tag{2.50}$$

If $m_1 = m_2$, the KL-divergence rate between $\{X_t\}$ and $\{Y_t\}$ is

$$\overline{D}_{KL}(\{X_t\}||\{Y_t\}) = \lim_{n \to \infty} \frac{1}{n} D_{KL}(X^n||Y^n) \tag{2.51}$$

If the PDF of the stochastic process $\{X_t\}$ is dependent on and continuously differentiable with respect to the parameter vector θ, then the Fisher information rate matrix (FIRM) is

$$\overline{J}_F(\theta) = \lim_{n \to \infty} \frac{1}{n} \int_{\mathbb{R}^{m_1 \times n}} \frac{1}{p(x^n, \theta)} \left[\frac{\partial}{\partial \theta} p(x^n, \theta) \right] \left[\frac{\partial}{\partial \theta} p(x^n, \theta) \right]^T dx^n \tag{2.52}$$

The information rates measure the average amount of information of stochastic processes in unit time. The limitations in Definition 2.6 may not exist, however, if the stochastic processes are stationary, these limitations in general exist. The following theorem gives the information rates for stationary Gaussian processes.

Theorem 2.7 Given two jointly Gaussian stationary processes $\{X_t \in \mathbb{R}^n, t \in Z\}$ and $\{Y_t \in \mathbb{R}^m, t \in Z\}$, with power spectral densities $S_X(\omega)$ and $S_Y(\omega)$, and $\{Z_t = [X_t^T, Y_t^T]^T \in \mathbb{R}^{n+m}, t \in Z\}$ with spectral density $S_Z(\omega)$, the entropy rate of the Gaussian process $\{X_t\}$ is

$$\overline{H}(\{X_t\}) = \frac{1}{2} \log(2\pi e)^n + \frac{1}{4\pi} \int_{-\pi}^{\pi} \log \det S_X(\omega) d\omega \tag{2.53}$$

The mutual information rate between $\{X_t\}$ and $\{Y_t\}$ is

$$\overline{I}(\{X_t\}; \{Y_t\}) = \frac{1}{4\pi} \int_{-\pi}^{\pi} \log \frac{\det S_X(\omega) \det S_Y(\omega)}{\det S_Z(\omega)} d\omega \tag{2.54}$$

If $m = n$, the KL-divergence rate between $\{X_t\}$ and $\{Y_t\}$ is

$$\overline{D}_{\mathrm{KL}}(\{X_t\}\|\{Y_t\}) = \frac{1}{4\pi} \int_{-\pi}^{\pi} \left\{ \log \frac{\det S_Y(\omega)}{\det S_X(\omega)} + Tr(S_Y^{-1}(\omega)(S_X(\omega) - S_Y(\omega))) \right\} d\omega$$

(2.55)

If the PDF of $\{X_t\}$ is dependent on and continuously differentiable with respect to the parameter vector θ, then the FIRM (assuming $n = 1$) is [163]

$$\overline{J}_{\mathrm{F}}(\theta) = \frac{1}{4\pi} \int_{-\pi}^{\pi} \frac{1}{S_X^2(\omega, \theta)} \left(\frac{\partial S_X(\omega, \theta)}{\partial \theta} \right) \left(\frac{\partial S_X(\omega, \theta)}{\partial \theta} \right)^T d\omega$$

(2.56)

Appendix B: α-Stable Distribution

α-stable distributions are a class of probability distributions satisfying the generalized central limit theorem, which are extensions of the Gaussian distribution. The Gaussian, inverse Gaussian, and Cauchy distributions are its special cases. Excepting the three kinds of distributions, other α-stable distributions do not have PDF with analytical expression. However, their characteristic functions can be written in the following form:

$$\Psi_X(\omega) = E[\exp(i\omega X)]$$
$$= \begin{cases} \exp[i\mu\omega - \gamma|\omega|^{\alpha}(1 + i\beta \, \mathrm{sign}(\omega)\tan(\pi\alpha/2))] & \text{for } \alpha \neq 1 \\ \exp[i\mu\omega - \gamma|\omega|^{\alpha}(1 + i\beta \, \mathrm{sign}(\omega)2\log|\omega|/\pi)] & \text{for } \alpha = 1 \end{cases}$$

(B.1)

where $\mu \in \mathbb{R}$ is the location parameter, $\gamma \geq 0$ is the dispersion parameter, $0 < \alpha \leq 2$ is the characteristic factor, $-1 \leq \beta \leq 1$ is the skewness factor. The parameter α determines the trailing of distribution. The smaller the value of α, the heavier the trail of the distribution is. The distribution is symmetric if $\beta = 0$, called the symmetric α-stable ($S\alpha S$) distribution. The Gaussian and Cauchy distributions are α-stable distributions with $\alpha = 2$ and $\alpha = 1$, respectively.

When $\alpha < 2$, the tail attenuation of α-stable distribution is slower than that of Gaussian distribution, which can be used to describe the outlier data or impulsive noises. In this case the distribution has infinite second-order moment, while the entropy is still finite.

Appendix C: Proof of (2.17)

Proof Assume ϕ is a concave function, and h is a monotonically increasing function. Denote h^{-1} the inverse of function h, we have

$$h^{-1}(H_{\phi}^h(X + Y)) = \int_{-\infty}^{+\infty} \phi[p_{X+Y}(\tau)]d\tau$$

(C.1)

Since X and Y are independent, then

$$p_{X+Y}(\tau) = \int_{-\infty}^{+\infty} p_X(t)p_Y(\tau - t)dt. \tag{C.2}$$

According to Jensen's inequality, we can derive

$$
\begin{aligned}
h^{-1}(H_\phi^h(X + Y)) &= \int_{-\infty}^{+\infty} \phi\left(\int_{-\infty}^{+\infty} p_Y(t)p_X(\tau - t)dt\right)d\tau \\
&\geq \int_{-\infty}^{+\infty} \left(\int_{-\infty}^{+\infty} p_Y(t)\phi[p_X(\tau - t)]dt\right)d\tau \\
&= \int_{-\infty}^{+\infty} p_Y(t)\left(\int_{-\infty}^{+\infty} \phi[p_X(\tau - t)]d\tau\right)dt \\
&= \int_{-\infty}^{+\infty} p_Y(t)(h^{-1}(H_\phi^h(X)))dt \\
&= h^{-1}(H_\phi^h(X))
\end{aligned}
\tag{C.3}
$$

As h is monotonically increasing, h^{-1} must also be monotonically increasing, thus we have $H_\phi^h(X + Y) \geq H_\phi^h(X)$. Similarly, $H_\phi^h(X + Y) \geq H_\phi^h(Y)$. Therefore,

$$H_\phi^h(X + Y) \geq \max\{H_\phi^h(X), H_\phi^h(Y)\} \tag{C.4}$$

For the case in which ϕ is a convex function and h is monotonically decreasing, the proof is similar (omitted).

Appendix D: Proof of Cramer–Rao Inequality

Proof First, one can derive the following two equalities:

$$
\begin{aligned}
E_{\theta_0}\left[\frac{\partial}{\partial\theta_0}\log p_t(Y, \theta_0)\right]^T &= \int_{\mathbb{R}^N}\left(\frac{\partial}{\partial\theta_0}\log p_Y(y, \theta_0)\right)^T p_Y(y, \theta_0)dy \\
&= \int_{\mathbb{R}^N}\left(\frac{\partial}{\partial\theta_0}p_Y(y, \theta_0)\right)^T dy \\
&= \left(\frac{\partial}{\partial\theta_0}\int_{\mathbb{R}^N}p_Y(y, \theta_0)dy\right)^T = \mathbf{0}
\end{aligned}
\tag{D.1}
$$

$$E_{\theta_0}\left\{\hat{\theta}(Y)\left[\frac{\partial}{\partial\theta_0}\log p_Y(Y,\theta_0)\right]^T\right\} = \int_{\mathbb{R}^N}\hat{\theta}(y)\left(\frac{\partial}{\partial\theta_0}\log p_Y(y,\theta_0)\right)^T p_Y(y,\theta_0)dy$$

$$= \int_{\mathbb{R}^N}\hat{\theta}(y)\left(\frac{\partial}{\partial\theta_0}p_Y(y,\theta_0)\right)^T dy$$

$$= \left(\frac{\partial}{\partial\theta_0}\int_{\mathbb{R}^N}\hat{\theta}(y)p_Y(y,\theta_0)dy\right)^T$$

$$= \left(\frac{\partial}{\partial\theta_0}E_{\theta_0}\left[\hat{\theta}(Y)\right]\right)^T = \left(\frac{\partial}{\partial\theta_0}\theta_0\right)^T = I$$

$$(D.2)$$

where I is a $d\times d$ identity matrix. Denote $\alpha = \hat{\theta}(Y) - \theta_0$ and $\beta = (\partial/\partial\theta_0)\log p_Y(Y,\theta_0)$.

Then

$$E_{\theta_0}[\alpha\beta^T] = E_{\theta_0}\left\{(\hat{\theta}(Y)-\theta_0)\left[\frac{\partial}{\partial\theta_0}\log p_Y(Y,\theta_0)\right]^T\right\}$$

$$= E_{\theta_0}\left\{\hat{\theta}(Y)\left[\frac{\partial}{\partial\theta_0}\log p_Y(Y,\theta_0)\right]^T\right\} - \theta_0 E_{\theta_0}\left[\frac{\partial}{\partial\theta_0}\log p_Y(Y,\theta_0)\right]^T$$

$$= I$$

$$(D.3)$$

So we obtain

$$E_{\theta_0}\begin{bmatrix}\alpha\\\beta\end{bmatrix}\begin{bmatrix}\alpha\\\beta\end{bmatrix}^T = \begin{bmatrix}E_{\theta_0}[\alpha\alpha^T] & I\\ I & E_{\theta_0}[\beta\beta^T]\end{bmatrix}\geq 0 \qquad (D.4)$$

According to the matrix theory, if the symmetric matrix $\begin{bmatrix}A & B\\ B^T & C\end{bmatrix}$ is positive-definite, then $A - BC^{-1}B^T \geq 0$. It follows that

$$E_{\theta_0}[\alpha\alpha^T] \geq (E_{\theta_0}[\beta\beta^T])^{-1} \qquad (D.5)$$

i.e., $P \geq J_F^{-1}(\theta_0)$.

3 Information Theoretic Parameter Estimation

Information theory is closely associated with the estimation theory. For example, the maximum entropy (MaxEnt) principle has been widely used to deal with estimation problems given incomplete knowledge or data. Another example is the Fisher information, which is a central concept in statistical estimation theory. Its inverse yields a fundamental lower bound on the variance of any unbiased estimator, i.e., the well-known Cramer–Rao lower bound (CRLB). An interesting link between information theory and estimation theory was also shown for the Gaussian channel, which relates the derivative of the mutual information with the minimum mean square error (MMSE) [81].

3.1 Traditional Methods for Parameter Estimation

Estimation theory is a branch of statistics and signal processing that deals with estimating the unknown values of parameters based on measured (observed) empirical data. Many estimation methods can be found in the literature. In general, the statistical estimation can be divided into two main categories: point estimation and interval estimation. The point estimation involves the use of empirical data to calculate a single value of an unknown parameter, while the interval estimation is the use of empirical data to calculate an interval of possible values of an unknown parameter. In this book, we only discuss the point estimation. The most common approaches to point estimation include the maximum likelihood (ML), method of moments (MM), MMSE (also known as Bayes least squared error), maximum *a posteriori* (MAP), and so on. These estimation methods also fall into two categories, namely, classical estimation (ML, MM, etc.) and Bayes estimation (MMSE, MAP, etc.).

3.1.1 Classical Estimation

The general description of the classical estimation is as follows: let the distribution function of population X be $F(x, \theta)$, where θ is an unknown (but deterministic) parameter that needs to be estimated. Suppose $X_1, X_2, \ldots, X_n$ are samples (usually independent and identically distributed, i.i.d.) coming from $F(x, \theta)$ ($x_1, x_2, \ldots, x_n$ are corresponding sample values). Then the goal of estimation is to construct an appropriate statistics $\hat{\theta}(X_1, X_2, \ldots, X_n)$ that serves as an approximation of unknown parameter θ. The statistics $\hat{\theta}(X_1, X_2, \ldots, X_n)$ is called an estimator of θ, and its

System Parameter Identification. DOI: http://dx.doi.org/10.1016/B978-0-12-404574-3.00003-8
© 2013 Tsinghua University Press Ltd. Published by Elsevier Inc. All rights reserved.

sample value $\hat{\theta}(x_1, x_2, \ldots, x_n)$ is called the estimated value of θ. Both the samples $\{X_i\}$ and the parameter θ can be vectors.

The ML estimation and the MM are two prevalent types of classical estimation.

3.1.1.1 ML Estimation

The ML method, proposed by the famous statistician R.A. Fisher, leads to many well-known estimation methods in statistics. The basic idea of ML method is quite simple: the event with greatest probability is most likely to occur. Thus, one should choose the parameter that maximizes the probability of the observed sample data. Assume that X is a continuous random variable with probability density function (PDF) $p(x, \theta)$, $\theta \in \Theta$, where θ is an unknown parameter, Θ is the set of all possible parameters. The ML estimate of parameter θ is expressed as

$$\hat{\theta} = \arg \max_{\theta \in \Theta} p(x_1, x_2, \ldots, x_n; \theta) \tag{3.1}$$

where $p(x_1, x_2, \ldots, x_n; \theta)$ is the joint PDF of samples $X_1, X_2, \ldots, X_n$. By considering the sample values $x_1, x_2, \ldots, x_n$ to be fixed "parameters," this joint PDF is a function of the parameter θ, called the likelihood function, denoted by $L(\theta)$. If samples $X_1, X_2, \ldots, X_n$ are i.i.d., we have $L(\theta) = \prod_{i=1}^{n} p(x_i, \theta)$. Then the ML estimate of θ becomes

$$\hat{\theta} = \arg \max_{\theta \in \Theta} L(\theta) = \arg \max_{\theta \in \Theta} \prod_{i=1}^{n} p(x_i, \theta) \tag{3.2}$$

In practice, it is often more convenient to work with the logarithm of the likelihood function (called the log-likelihood function). In this case, we have

$$\hat{\theta} = \arg \max_{\theta \in \Theta} \{\log L(\theta)\} = \arg \max_{\theta \in \Theta} \left\{ \sum_{i=1}^{n} \log p(x_i, \theta) \right\} \tag{3.3}$$

An ML estimate is the same regardless of whether we maximize the likelihood or log-likelihood function, since log is a monotone transformation.

In most cases, the ML estimate can be solved by setting the derivative of the log-likelihood function to zero:

$$\frac{\partial \log L(\theta)}{\partial \theta} = 0 \tag{3.4}$$

For many models, however, there is no closed form solution of ML estimate, and it has to be found numerically using optimization methods.

If the likelihood function involves latent variables in addition to unknown parameter θ and known data observations $x_1, x_2, \ldots, x_n$, one can use the expectation−maximization (EM) algorithm to find the ML solution [164,165] (see Appendix E). Typically, the latent variables are included in a likelihood function because either there are missing values among the data or the model can be formulated more simply by assuming the existence of additional unobserved data points.

ML estimators possess a number of attractive properties especially when sample size tends to infinity. In general, they have the following properties:

- Consistency: As the sample size increases, the estimator converges in probability to the true value being estimated.
- Asymptotic normality: As the sample size increases, the distribution of the estimator tends to the Gaussian distribution.
- Efficiency: The estimator achieves the CRLB when the sample size tends to infinity.

3.1.1.2 Method of Moments

The MM uses the sample algebraic moments to approximate the population algebraic moments, and then solves the parameters. Consider a continuous random variable X, with PDF $p(x, \theta_1, \theta_2, \ldots, \theta_k)$, where $\theta_1, \theta_2, \ldots, \theta_k$ are k unknown parameters. By the law of large numbers, the l-order sample moment $A_l = (1/n) \sum_{i=1}^{n} X_i^l$ of X will converge in probability to the l-order population moment $\mu_l = E(X^l)$, which is a function of $(\theta_1, \theta_2, \ldots, \theta_k)$, i.e.,

$$A_l \xrightarrow[n \to \infty]{p} \mu_l(\theta_1, \theta_2, \ldots, \theta_k) \quad l = 1, 2, \ldots \tag{3.5}$$

The sample moment A_l is a good approximation of the population moment μ_l, thus one can achieve an estimator of parameters θ_i $(i = 1, 2, \ldots, k)$ by solving the following equations:

$$\begin{cases} A_1 &= \mu_1(\theta_1, \theta_2, \ldots, \theta_k) \\ A_2 &= \mu_2(\theta_1, \theta_2, \ldots, \theta_k) \\ &\vdots \\ A_k &= \mu_k(\theta_1, \theta_2, \ldots, \theta_k) \end{cases} \tag{3.6}$$

The solution of (3.6) is the MM estimator, denoted by $\hat{\theta}_i(A_1, A_2, \ldots, A_k)$, $i = 1, 2, \ldots, k$.

3.1.2 Bayes Estimation

The basic viewpoint of Bayes statistics is that in any statistic reasoning problem, a prior distribution must be prescribed as a basic factor in the reasoning process, besides the availability of empirical data. Unlike classical estimation, the Bayes estimation regards the unknown parameter as a random variable (or random vector) with some prior distribution. In many situations, this prior distribution does not need to be precise, which can be even improper (e.g., uniform distribution on the whole space). Since the unknown parameter is a random variable, in the following we use X to denote the parameter to be estimated, and Y to denote the observation data $Y = [Y_1, Y_2, \ldots, Y_n]^T$.

Assume that both the parameter X and observation Y are continuous random variables with joint PDF

$$p(x, y) = p(x).p(y|x) \tag{3.7}$$

where $p(x)$ is the marginal PDF of X (the prior PDF) and $p(y|x)$ is the conditional PDF of Y given $X = x$ (also known as the likelihood function if considering x as the function's variable). By using the Bayes formula, one can obtain the posterior PDF of X given $Y = y$:

$$p(x|y) = \frac{p(y|x)p(x)}{\int p(y|x)p(x)dx} \tag{3.8}$$

Let $\hat{X} = g(Y)$ be an estimator of X (based on the observation Y), and let $l(X, \hat{X})$ be a loss function that measures the difference between random variables X and $\hat{X}$. The Bayes risk of $\hat{X}$ is defined as the expected loss (the expectation is taken over the joint distribution of X and Y):

$$
\begin{aligned}
R(X, \hat{X}) &= E[l(X, \hat{X})] \\
&= \int l(x, \hat{x})p(x, y)dx\, dy \\
&= \int \left(\int l(x, \hat{x})p(x|y)dx \right)p(y)dy \\
&= \int R(X, \hat{X}|y)p(y)dy
\end{aligned}
\tag{3.9}
$$

where $R(X, \hat{X}|y)$ denotes the posterior expected loss (posterior Bayes risk). An estimator is said to be a Bayes estimator if it minimizes the Bayes risk among all estimators. Thus, the Bayes estimator can be obtained by solving the following optimization problem:

$$g^* = \arg \min_{g \in G} R(X, \hat{X}) \tag{3.10}$$

where G denotes all Borel measurable functions $g : y \mapsto \hat{x}$. Obviously, the Bayes estimator also minimizes the posterior Bayes risk for each y.

The loss function in Bayes risk is usually a function of the estimation error $e = X - \hat{X}$. The common loss functions used for Bayes estimation include:

1. squared error function: $l(e) = e^2$;
2. absolute error function: $l(e) = |e|$;
3. 0−1 loss function: $l(e) = 1 - \delta(e)$, where $\delta(.)$ denotes the delta function.[1]

---

[1] For a discrete variable x, $\delta(x)$ is defined by $\delta(x) = \begin{cases} 1, & \text{if } x = 0 \\ 0, & \text{if } x \neq 0 \end{cases}$, while for a continuous variable, it is defined as $\delta(x) = \begin{cases} \infty, & \text{if } x = 0 \\ 0, & \text{if } x \neq 0 \end{cases}$, satisfying $\int \delta(x)dx = 1$.

The squared error loss corresponds to the MSE criterion, which is perhaps the most prevalent risk function in use due to its simplicity and efficiency. With the above loss functions, the Bayes estimates of the unknown parameter are, respectively, the mean, median, and mode[2] of the posterior PDF $p(x|y)$, i.e.,

$$\begin{cases} (a) & \hat{x} = \int x p(x|y) dx = E(X|y) \\ (b) & \int_{-\infty}^{\hat{x}} p(x|y) dx = \int_{\hat{x}}^{+\infty} p(x|y) dx \\ (c) & \hat{x} = \arg \max_{x} p(x|y) \end{cases} \tag{3.11}$$

The estimators (a) and (c) in (3.11) are known as the MMSE and MAP estimators. A simple proof of the MMSE estimator is given in Appendix F. It should be noted that if the posterior PDF is symmetric and unimodal (SUM, such as Gaussian distribution), the three Bayes estimators are identical.

The MAP estimate is a mode of the posterior distribution. It is a limit of Bayes estimation under $0-1$ loss function. When the prior distribution is uniform (i.e., a constant function), the MAP estimation coincides with the ML estimation. Actually, in this case we have

$$\begin{aligned} \hat{x}_{\text{MAP}} &= \arg \max_{x} p(x|y) \\ &= \arg \max_{x} \frac{p(y|x)p(x)}{\int p(y|x)p(x) dx} \\ &= \arg \max_{x} p(y|x)p(x) \\ &\stackrel{(a)}{=} \arg \max_{x} p(y|x) = \hat{x}_{\text{ML}} \end{aligned} \tag{3.12}$$

where (a) comes from the fact that $p(x)$ is a constant.

Besides the previous common risks, other Bayes risks can be conceived. Important examples include the mean p-power error [30], Huber's M-estimation cost [33], and the risk-sensitive cost [38], etc. It has been shown in [24] that if the posterior PDF is symmetric, the posterior mean is an optimal estimate for a large family of Bayes risks, where the loss function is even and convex.

In general, a Bayes estimator is a nonlinear function of the observation. However, if X and Y are jointly Gaussian, then the MMSE estimator is linear. Suppose $X \in \mathbb{R}^m$, $Y \in \mathbb{R}^n$, with jointly Gaussian PDF

$$p(x, y) = (2\pi)^{-(m+n)/2} (\det C)^{-1/2} \exp\left\{ -\frac{1}{2} \begin{bmatrix} x - E(X) \\ y - E(Y) \end{bmatrix}^T C^{-1} \begin{bmatrix} x - E(X) \\ y - E(Y) \end{bmatrix} \right\} \tag{3.13}$$

---

[2] The mode of a continuous probability distribution is the value at which its PDF attains its maximum value.

where C is the covariance matrix:

$$C = \begin{bmatrix} C_{XX} & C_{XY} \\ C_{YX} & C_{YY} \end{bmatrix} \tag{3.14}$$

Then the posterior PDF $p(x|y)$ is also Gaussian and has mean (the MMSE estimate)

$$E(X|y) = E(X) + C_{XY}C_{YY}^{-1}(y - E(Y)) \tag{3.15}$$

which is, obviously, a linear function of y.

There are close relationships between estimation theory and information theory. The concepts and principles in information theory can throw new light on estimation problems and suggest new methods for parameter estimation. In the sequel, we will discuss information theoretic approaches to parameter estimation.

3.2 Information Theoretic Approaches to Classical Estimation

In the literature, there have been many reports on the use of information theory to deal with classical estimation problems (e.g., see [149]). Here, we only give several typical examples.

3.2.1 Entropy Matching Method

Similar to the MM, the entropy matching method obtains the parameter estimator by using the sample entropy (entropy estimator) to approximate the population entropy. Suppose the PDF of population X is $p(x, \theta)$ (θ is an unknown parameter). Then its differential entropy is

$$H(\theta) = -\int p(x, \theta)\log p(x, \theta)dx \tag{3.16}$$

At the same time, one can use the sample $(X_1, X_2, \ldots, X_n)$ to calculate the sample entropy $\hat{H}(X_1, X_2, \ldots, X_n)$.[3] Thus, we can obtain an estimator of parameter θ through solving the following equation:

$$H(\theta) = \hat{H}(X_1, X_2, \ldots, X_n) \tag{3.17}$$

If there are several parameters, the above equation may have infinite number of solutions, while a unique solution can be achieved by combining the MM. In [166], the entropy matching method was used to estimate parameters of generalized Gaussian distribution (GGD).

---

[3] Several entropy estimation methods will be presented in Chapter 4.

3.2.2   Maximum Entropy Method

The maximum entropy method applies the famous MaxEnt principle to parameter estimation. The basic idea is that, subject to the information available, one should choose the parameter θ such that the entropy is as large as possible, or the distribution as nearly uniform as possible. Here, the maximum entropy method refers to a general approach rather than a specific parameter estimation method. In the following, we give three examples of maximum entropy method.

3.2.2.1   Parameter Estimation of Exponential Type Distribution

Assume that the PDF of population X is of the following form:

$$p(x, \theta) = \exp\left(-\theta_0 - \sum_{k=1}^{K} \theta_k g_k(x) \right) \tag{3.18}$$

where $g_k(x)$, $k = 1, \ldots, K$, are K (generalized) characteristic moment functions, $\theta = (\theta_0, \theta_1, \ldots, \theta_K)$ is an unknown parameter vector to be estimated. Many known probability distributions are special cases of this exponential type distribution. By Theorem 2.2, $p(x, \theta)$ is the maximum entropy distribution satisfying the following constraints:

$$\begin{cases} \int_{\mathbb{R}} p(x)\mathrm{d}x = 1 \\ \int_{\mathbb{R}} g_k(x)p(x)\mathrm{d}x = \mu_k(\theta), \quad k = 1, 2, \ldots, K \end{cases} \tag{3.19}$$

where $\mu_k(\theta)$ denotes the population characteristic moment:

$$\mu_k(\theta) = \int_{\mathbb{R}} g_k(x)p(x, \theta)\mathrm{d}x \tag{3.20}$$

As θ is unknown, the population characteristic moments cannot be calculated. We can approximate them using the sample characteristic moments. And then, an estimator of parameter θ can be obtained by solving the following optimization problem:

$$\begin{cases} \max_{p} H(p) = -\int_{\mathbb{R}} p(x)\log p(x)\mathrm{d}x \\ \text{s.t.} \begin{cases} \int_{\mathbb{R}} p(x)\mathrm{d}x = 1 \\ \int_{\mathbb{R}} g_k(x)p(x)\mathrm{d}x = \hat{\mu}_k, \quad k = 1, 2, \ldots, K \end{cases} \end{cases} \tag{3.21}$$

where $\hat{\mu}_k$, $k = 1, \ldots, K$, are K sample characteristic moments, i.e.,

$$\hat{\mu}_k = \frac{1}{n}\sum_{i=1}^{n} g_k(X_i), \quad k = 1, 2, \ldots, K \tag{3.22}$$

According to Theorem 2.2, the estimator of θ satisfies the equations:

$$\begin{cases} \int_{\mathbb{R}} \exp\left(-\sum_{k=1}^{K} \hat{\theta}_k g_k(x)\right) dx = \exp(\hat{\theta}_0) \\[2em] \dfrac{\int_{\mathbb{R}} g_i(x)\exp\left(-\sum_{k=1}^{K} \hat{\theta}_k g_k(x)\right) dx}{\exp(\hat{\theta}_0)} = \hat{\mu}_i, \quad i = 1, 2, \ldots, K \end{cases} \tag{3.23}$$

If $g_k(x) = x^k$, the above estimation method will be equivalent to the MM.

3.2.2.2 Maximum Spacing Estimation

Suppose the distribution function of population X is $F(x, \theta)$, and the true value of the unknown parameter θ is θ_0, then the random variable $X^* = F(X, \theta)$ will be distributed over the interval $[0, 1]$, which is a uniform distribution if $\theta = \theta_0$. According to the MaxEnt principle, if the distribution over a finite interval is uniform, the entropy will achieve its maximum. Therefore, the entropy of random variable X^* will attain the maximum value if $\theta = \theta_0$. So one can obtain an estimator of the parameter θ by maximizing the sample entropy of X^*. Let a sample of population X be $(X_1, X_2, \ldots, X_n)$, the sample of X^* will be $(F(X_1, \theta), F(X_2, \theta), \ldots, F(X_n, \theta))$. Let $\hat{H}(F(X_1, \theta), F(X_2, \theta), \ldots, F(X_n, \theta))$ denote the sample entropy of X^*, the estimator of parameter θ can be expressed as

$$\hat{\theta} = \arg\max_{\theta} \hat{H}(F(X_1, \theta), F(X_2, \theta), \ldots, F(X_n, \theta)) \tag{3.24}$$

If the sample entropy is calculated by using the one-spacing estimation method (see Chapter 4), then we have

$$\hat{\theta} = \arg\max_{\theta} \sum_{i=1}^{n-1} \log\{F(X_{n,i+1}, \theta) - F(X_{n,i}, \theta)\} \tag{3.25}$$

where $(X_{n,1}, X_{n,2}, \ldots, X_{n,n})$ is the order statistics of $(X_1, X_2, \ldots, X_n)$. Formula (3.25) is called the maximum spacing estimation of parameter θ.

3.2.2.3 Maximum Equality Estimation

Suppose $(X_1, X_2, \ldots, X_n)$ is an i.i.d. sample of population X with PDF $p(x, \theta)$. Let $X_{n,1} \leq X_{n,2} \leq \cdots \leq X_{n,n}$ be the order statistics of $(X_1, X_2, \ldots, X_n)$. Then the random sample divides the real axis into $n+1$ subintervals $(X_{n,i}, X_{n,i+1})$, $i = 0, 1, \ldots, n$, where $X_{n,0} = -\infty$ and $X_{n,n+1} = +\infty$. Each subinterval has the probability:

$$P_i = \int_{X_{n,i}}^{X_{n,i+1}} p(x, \theta) \mathrm{d}x, \quad i = 0, 1, \ldots, n \tag{3.26}$$

Since the sample is random and i.i.d., the most reasonable situation is that the probabilities of $n+1$ subinterval are equal. Hence, the parameter θ should be chosen in such a way as to maximize the entropy of distribution $\{P_i\}$ (or to make $\{P_i\}$ as nearly uniform as possible), i.e.,

$$\begin{aligned}
\hat{\theta} &= \arg\max_{\theta} \left\{ -\sum_{i=0}^{n} P_i \log P_i \right\} \\
&= \arg\max_{\theta} \left\{ -\sum_{i=0}^{n} \left(\int_{X_{n,i}}^{X_{n,i+1}} p(x, \theta) \mathrm{d}x \right) \log \left(\int_{X_{n,i}}^{X_{n,i+1}} p(x, \theta) \mathrm{d}x \right) \right\}
\end{aligned} \tag{3.27}$$

The above estimation is called the maximum equality estimation of parameter θ.

It is worth noting that besides parameter estimation, the MaxEnt principle can also be applied to spectral density estimation [48]. The general idea is that the maximum entropy rate stochastic process that satisfies the given constant autocorrelation and variance constraints, is a linear Gauss−Markov process with i.i.d. zero-mean, Gaussian input.

3.2.3 Minimum Divergence Estimation

Let $(X_1, X_2, \ldots, X_n)$ be an i.i.d. random sample from a population X with PDF $p(x, \theta)$, $\theta \in \Theta$. Let $\hat{p}_n(x)$ be the estimated PDF based on the sample. Let $\hat{\theta}$ be an estimator of θ. Then $p(x, \hat{\theta})$ is also an estimator for $p(x, \theta)$. Then the estimator $\hat{\theta}$ should be chosen so that $p(x, \hat{\theta})$ is as close as possible to $\hat{p}_n(x)$. This can be achieved by minimizing any measure of information divergence, say the KL-divergence

$$\begin{aligned}
\hat{\theta} &= \arg\min_{\theta} D_{\mathrm{KL}}(\hat{p}_n(x) \| p(x, \theta)) \\
&= \arg\min_{\theta} \int \hat{p}_n(x) \log \frac{\hat{p}_n(x)}{p(x, \theta)} \mathrm{d}x
\end{aligned} \tag{3.28}$$

or, alternatively,

$$\begin{aligned}
\hat{\theta} &= \arg\min_{\theta} D_{\mathrm{KL}}(p(x, \theta) \| \hat{p}_n(x)) \\
&= \arg\min_{\theta} \int p(x, \theta) \log \frac{p(x, \theta)}{\hat{p}_n(x)} \mathrm{d}x
\end{aligned} \tag{3.29}$$

The estimate $\hat{\theta}$ in (3.28) or (3.29) is called the minimum divergence (MD) estimate of θ. In practice, we usually use (3.28) for parameter estimation, because it can be simplified as

$$\hat{\theta} = \arg\max_{\theta} \int \hat{p}_n(x)\log p(x, \theta)\mathrm{d}x \tag{3.30}$$

Depending on the estimated PDF $\hat{p}_n(x)$, the MD estimator may take many different forms. Next, we present three specific examples of MD estimator.

- **MD Estimator 1**

   Without loss of generality, we assume that the sample satisfies $x_1 < x_2 < \cdots < x_n$. Then the distribution function can be estimated as

$$\hat{F}_n(x) = \begin{cases} 0 & x < x_1 \\ 1/n & x_1 \le x < x_2 \\ 2/n & x_2 \le x < x_3 \\ \vdots & \\ 1 & x \ge x_n \end{cases} \tag{3.31}$$

   Thus, we have

$$\begin{aligned} \hat{\theta} &= \arg\max_{\theta} \int \hat{p}_n(x)\log p(x, \theta)\mathrm{d}x \\ &= \arg\max_{\theta} \int \log p(x, \theta)\mathrm{d}\hat{F}_n(x) \\ &= \arg\max_{\theta} \frac{1}{n}\sum_{i=1}^{n}\log p(x_i, \theta) \\ &= \arg\max_{\theta} \log L(\theta) \end{aligned} \tag{3.32}$$

   where $L(\theta) = \prod_{i=1}^{n} p(x_i, \theta)$ is the likelihood function. In this case, the MD estimation is exactly the ML estimation.

- **MD Estimator 2**

   Suppose the population X is distributed over the interval $[x_0, x_{n+1}]$, and the sample satisfies $x_1 < x_2 < \cdots < x_n$. It is reasonable to assume that in each subinterval $[x_i, x_{i+1}]$, the probability is $1/(n + 1)$. And hence, the PDF of X can be estimated as

$$\hat{p}_n(x) = \frac{1}{(n + 1)(x_{i+1} - x_i)} \quad \text{if } x_i \le x < x_{i+1} \ (i = 0, 1, \ldots, n) \tag{3.33}$$

   Substituting (3.33) into (3.30) yields

$$\begin{aligned} \hat{\theta} &= \arg\max_{\theta} \sum_{i=0}^{n} \int_{x_i}^{x_{i+1}} \frac{1}{(n + 1)(x_{i+1} - x_i)}\log p(x, \theta)\mathrm{d}x \\ &= \arg\max_{\theta} \sum_{i=0}^{n} (x_{i+1} - x_i)^{-1} \int_{x_i}^{x_{i+1}} \log p(x, \theta)\mathrm{d}x \end{aligned} \tag{3.34}$$

If $p(x, \theta)$ is a continuous function of x, then according to the mean value theorem of integral calculus, we have

$$(x_{i+1} - x_i)^{-1} \int_{x_i}^{x_{i+1}} \log p(x, \theta) dx = \log p(\bar{x}_i, \theta) \qquad (3.35)$$

where $\bar{x}_i \in [x_i, x_{i+1}]$. Hence, (3.34) can be written as

$$\hat{\theta} = \arg\max_{\theta} \sum_{i=0}^{n} \log p(\bar{x}_i, \theta) \qquad (3.36)$$

The above parameter estimation is in form very similar to the ML estimation. However, different from Fisher's likelihood function, the values of the cost function in (3.36) are taken at $\bar{x}_0, \bar{x}_1, \ldots, \bar{x}_n$, which are determined by mean value theorem of integral calculus.

- **MD Estimator 3**

Assume population X is distributed over the interval $[d_0, d_{m+1}]$, and this interval is divided into $m + 1$ subintervals $[d_i, d_{i+1}]$ $(i = 0, 1, \ldots, m)$ by m data $d_1 < d_2 < \cdots < d_n$. The probability of each subinterval is determined by

$$P_i = \int_{d_i}^{d_{i+1}} p(x, \theta) dx, \quad i = 0, 1, \ldots, m \qquad (3.37)$$

If $Q_0, Q_1, \ldots, Q_m$ are given proportions of the population that lie in the $m + 1$ subintervals ($\sum_{i=0}^{m} Q_i = 1$), then parameter θ should be chosen so as to make $\{P_i\}$ and $\{Q_i\}$ as close as possible, i.e.,

$$
\begin{aligned}
\hat{\theta} &= \arg\min_{\theta} D_{\mathrm{KL}}(Q\|P) \\
&= \arg\min_{\theta} \sum_{i=0}^{m} Q_i \log \frac{Q_i}{P_i} \\
&= \arg\max_{\theta} \sum_{i=0}^{m} Q_i \log \int_{d_i}^{d_{i+1}} p(x, \theta) dx
\end{aligned}
\qquad (3.38)
$$

This is a useful estimation approach, especially when the information available is on proportions in the population, such as proportions of persons in different income intervals or proportions of students in different score intervals.

In the previous MD estimations, the KL-divergence can be substituted by other definitions of divergence. For instance, if using ϕ-divergence, we have

$$\hat{\theta} = \arg\min_{\theta} D_{\phi}(\hat{p}_n(x)\|p(x, \theta)) \qquad (3.39)$$

or

$$\hat{\theta} = \arg\min_{\theta} D_{\phi}(p(x, \theta)\|\hat{p}_n(x)) \qquad (3.40)$$

For details on the minimum ϕ-divergence estimation, the readers can refer to [130].

3.3 Information Theoretic Approaches to Bayes Estimation

The Bayes estimation can also be embedded within the framework of information theory. In particular, some information theoretic measures, such as the entropy and correntropy, can be used instead of the traditional Bayes risks.

3.3.1 Minimum Error Entropy Estimation

In the scenario of Bayes estimation, the minimum error entropy (MEE) estimation aims to minimize the entropy of the estimation error, and hence decrease the uncertainty in estimation. Given two random variables: $X \in \mathbb{R}^m$, an unknown parameter to be estimated, and $Y \in \mathbb{R}^n$, the observation (or measurement), the MEE (with Shannon entropy) estimation of X based on Y can be formulated as

$$
\begin{aligned}
g_{\mathrm{MEE}} &= \arg\min_{g \in G} H(e) \\
&= \arg\min_{g \in G} H(X - g(Y)) \\
&= \arg\min_{g \in G} -\int_{\mathbb{R}^m} p_e(\xi)\log p_e(\xi)\mathrm{d}\xi
\end{aligned}
\tag{3.41}
$$

where $e = X - g(Y)$ is the estimation error, $g(Y)$ is an estimator of X based on Y, g is a measurable function, G stands for the collection of all measurable functions $g:\mathbb{R}^n \to \mathbb{R}^m$, and $p_e(.)$ denotes the PDF of the estimation error. When (3.41) is compared with (3.9) one concludes that the "loss function" in MEE is $-\log p_e(.)$, which is different from traditional Bayesian risks, like MSE. Indeed one does not need to impose a risk functional in MEE, the risk is directly related to the error PDF. Obviously, other entropy definitions (such as order-α Renyi entropy) can also be used in MEE estimation. This feature is potentially beneficial because the risk is matched to the error distribution.

The early work in MEE estimation can be traced back to the late 1960s when Weidemann and Stear [86] studied the use of error entropy as a criterion function (risk function) for analyzing the performance of sampled data estimation systems. They proved that minimizing the error entropy is equivalent to minimizing the mutual information between the error and the observation, and also proved that the reduced error entropy is upper-bounded by the amount of information obtained by the observation. Minamide [89] extended Weidemann and Stear's results to a continuous-time estimation system. Tomita et al. applied the MEE criterion to linear Gaussian systems and studied state estimation (Kalman filtering), smoothing, and predicting problems from the information theory viewpoint. In recent years, the MEE became an important criterion in supervised machine learning [64].

In the following, we present some important properties of the MEE criterion, and discuss its relationship to conventional Bayes risks. For simplicity, we assume that the error e is a scalar ($m = 1$). The extension to arbitrary dimensions will be straightforward.

3.3.1.1 Some Properties of MEE Criterion

Property 1: $\forall c \in \mathbb{R}, H(e + c) = H(e)$.

Proof: This is the shift invariance of the differential entropy. According to the definition of differential entropy, we have

$$
\begin{aligned}
H(e + c) &= - \int p_{e+c}(\xi)\log p_{e+c}(\xi)\mathrm{d}\xi \\
&= - \int p_e(\xi - c)\log p_e(\xi - c)\mathrm{d}\xi \\
&= - \int p_e(\xi)\log p_e(\xi)\mathrm{d}\xi = H(e)
\end{aligned}
\tag{3.42}
$$

Remark: The MEE criterion is invariant with respect to error's mean. In practice, in order to meet the desire for small error values, the MEE estimate is usually restricted to zero-mean (unbiased) error, which requires special user attention (i.e., mean removal). We should note that the unbiased MEE estimate can still be non-unique (see Property 6).

Property 2: If ζ is a random variable independent of the error e, then $H(e + \zeta) \geq H(e)$.

Proof: According to the properties of differential entropy and the independence condition, we have

$$
H(e + \zeta) \geq H(e + \zeta | \zeta) = H(e | \zeta) = H(e)
\tag{3.43}
$$

Remark: Property 2 implies that MEE criterion is robust to independent additive noise. Specifically, if error e contains an independent additive noise ζ, i.e., $e = e_T + \zeta$, where e_T is the true error, then minimizing the contaminated error entropy $H(e)$ will constrain the true error entropy $H(e_T)$ ($H(e_T) \leq H(e)$).

Property 3: Minimizing the error entropy $H(e)$ is equivalent to minimizing the mutual information between error e and the observation Y, i.e., $\min_{g \in G} H(e) \Leftrightarrow \min_{g \in G} I(e; Y)$.

Proof: As $I(e; Y) = H(e) - H(e|Y)$, we have

$$H(e) = I(e; Y) + H(e|Y)$$
$$\qquad = I(e; Y) + H(X - g(Y)|Y) \qquad\qquad\qquad (3.44)$$
$$\qquad = I(e; Y) + H(X|Y)$$

It follows easily that

$$\min_{g \in G} H(e) \Leftrightarrow \min_{g \in G}\{I(e; Y) + H(X|Y)\}$$
$$\overset{(a)}{\Leftrightarrow} \min_{g \in G} I(e; Y) \qquad\qquad\qquad (3.45)$$

where (a) comes from the fact that the conditional entropy $H(X|Y)$ is not related to g.

Remark: The minimization of error entropy will minimize the mutual information between the error and the observation. Hence, under MEE criterion, the observation will be "fully utilized," so that the error contains least information about the observation.

Property 4: The error entropy $H(e)$ is lower bounded by the conditional entropy $H(X|Y)$, and this lower bound is achieved if and only if error e is independent of the observation Y.

Proof: By (3.44), we have

$$H(e) = I(e; Y) + H(X|Y) \geq H(X|Y) \qquad\qquad\qquad (3.46)$$

where the inequality follows from the fact that $I(e; Y) \geq 0$, with equality if and only if error e and observation Y are independent.

Remark: The error entropy can never be smaller than the conditional entropy of the parameter X given observation Y. This lower bound is achieved if and only if the error contains no information about the observation.

For MEE estimation, there is no explicit expression for the optimal estimate unless some constraints on the posterior PDF are imposed. The next property shows that, if the posterior PDF is SUM, the MEE estimate will be equal to the conditional mean (i.e., the MMSE estimate).

Property 5: If for any y, the posterior PDF $p(x|y)$ is SUM, then the MEE estimate equals the posterior mean, i.e., $g_{\mathrm{MEE}}(y) = E(X|Y = y)$.

Proof: This property is a direct consequence of Theorem 1 of [91] (Omitted).

Remark: Since the posterior PDF is SUM, in the above property the MEE estimate also equals the posterior median or posterior mode. We point out that for order-α

Table 3.1 Optimal Estimates for Several Bayes Estimation Methods

Bayes Estimation	Risk Function	Optimal Estimate
MMSE	$\int_{\mathbb{R}} \xi^2 p_e(\xi) d\xi$	$g_{\text{MMSE}} = \text{mean}[p(.|y)]$
Mean absolute deviation (MAD)	$\int_{\mathbb{R}} |\xi| p_e(\xi) d\xi$	$g_{\text{MAD}} = \text{median}[p(.|y)]$
MAP	$\int_{\mathbb{R}} l_{0-1}(\xi) p_e(\xi) d\xi$	$g_{\text{MAP}} = \text{mode}[p(.|y)]$
General Bayes estimation (GBE)	$\int_{\mathbb{R}} l(\xi) p_e(\xi) d\xi$	If $l(.)$ is even and convex, and $\forall y$, $p(x|y)$ is symmetric in x, then $g_{\text{GBE}}(.) = \text{mean}[p(.|y)]$
MEE	$-\int_{\mathbb{R}} p_e(\xi) \log p_e(\xi) d\xi$	If $\forall y$, $p(x|y)$ is SUM, then $g_{\text{MEE}}(.) = \text{mean}[p(.|y)]$

Renyi entropy, this property still holds. One may be led to conclude that MEE has no advantage over the MMSE, since they correspond to the same solution for SUM case. However, the risk functional in MEE is still well defined for unimodal but asymmetric distributions, although no general results are known for this class of distributions. But we suspect that the practical advantages of MEE versus MSE reported in the literature [64] are taking advantage of this case. In the context of adaptive filtering, the two criteria may yield much different *performance surfaces* even if they have the same optimal solution.

Table 3.1 gives a summary of the optimal estimates for several Bayes estimation methods.

Property 6: The MEE estimate may be nonunique even if the error distribution is restricted to zero mean (unbiased).

Proof: We prove this property by means of a simple example as follows [94]: suppose Y is a discrete random variable with Bernoulli distribution:

$$\Pr(Y = 0) = \Pr(Y = 1) = 0.5 \tag{3.47}$$

The posterior PDF of X given Y is $(a > 0)$:

$$p(x|Y = 0) = \begin{cases} \dfrac{1}{2a} & \text{if } |x| \le a \\ 0 & \text{other} \end{cases}$$

$$p(x|Y = 1) = \begin{cases} \dfrac{1}{a} & \text{if } |x| \le \dfrac{1}{2}a \\ 0 & \text{other} \end{cases}$$

Given an estimator $\hat{X} = g(Y)$, the error PDF will be

$$p_e(x) = \frac{1}{2}\{p(x + g(0)|Y = 0) + p(x + g(1)|Y = 1)\} \tag{3.48}$$

Let g be an unbiased estimator, then $\int_{-\infty}^{\infty} x p_e(x)dx = 0$, and hence $g(0) = -g(1)$. Assuming $g(0) \geq 0$ (due to symmetry, one can obtain similar results for $g(0) < 0$), the error entropy can be calculated as

$$H(e) = \begin{cases} -\frac{3}{4}\log(3) + \log(4a), & \text{if } 0 \leq g(0) \leq \frac{1}{4}a \\[2mm] \left(-\frac{9}{8} + \frac{3g(0)}{2a}\right)\log(3) + \left(\frac{1}{4} - \frac{g(0)}{a}\right)\log(2) + \log(4a), & \text{if } \frac{1}{4}a < g(0) \leq \frac{3}{4}a \\[2mm] -\frac{1}{2}\log(2) + \log(4a), & \text{if } g(0) > \frac{3}{4}a \end{cases}$$

$$\tag{3.49}$$

One can easily verify that the error entropy achieves its minimum value when $0 \leq g(0) \leq a/4$. Clearly, in this example there are infinitely many unbiased MEE estimators.

Property 7 (Score Orthogonality [92]): Given an MEE estimator $\hat{X} = g_{\text{MEE}}(Y)$, the error's score $\psi(e|g_{\text{MEE}})$ is orthogonal to any measurable function $\varphi(Y)$ of Y, where $\psi(e|g_{\text{MEE}})$ is defined as

$$\psi(e|g_{\text{MEE}}) = \frac{\partial}{\partial e}\{\log p(e|g_{\text{MEE}})\} \tag{3.50}$$

where $p(e|g)$ denotes the error PDF for estimator $\hat{X} = g(Y)$.

Proof: Given an MEE estimator $g_{\text{MEE}} \in G$, $\forall \varphi \in G$ and $\forall \gamma \in \mathbb{R}$, we have

$$H(e|g_{\text{MEE}}) = \underset{g \in G}{\arg \min} H(e) \leq H(e|g_{\text{MEE}} + \gamma\varphi) = H(e - \gamma\varphi(Y)|g_{\text{MEE}}) \tag{3.51}$$

For $|\gamma|$ small enough, $\gamma\varphi(Y)$ will be a "small" random variable. According to [167], we have

$$H(e - \gamma\varphi(Y)|g_{\text{MEE}}) - H(e|g_{\text{MEE}}) = \gamma E[\psi(e|g_{\text{MEE}})\varphi(Y)] + o(\gamma\varphi(Y)) \tag{3.52}$$

where $o(.)$ denotes the higher order terms. Then the derivative of entropy $H(e - \gamma\varphi(Y)|g_{\text{MEE}})$ with respect to γ at $\gamma = 0$ is

$$\frac{\partial}{\partial \gamma}\{H(e - \gamma\varphi(Y)|g_{\text{MEE}})\}|_{\gamma=0} = E[\psi(e|g_{\text{MEE}})\varphi(Y)] \tag{3.53}$$

Combining (3.51) and (3.53) yields

$$E[\psi(e|g_{\text{MEE}})\varphi(Y)] = 0, \quad \forall\, \varphi \in G \tag{3.54}$$

Remark: If the error is zero-mean Gaussian distributed with variance σ^2, the score function will be $\psi(e|g_{\text{MEE}}) = -e/\sigma^2$. In this case, the score orthogonality condition reduces to $E[e\varphi(Y)] = 0$. This is the well-known orthogonality condition for MMSE estimation.

In MMSE estimation, the orthogonality condition is a *necessary and sufficient* condition for optimality, and can be used to find the MMSE estimator. In MEE estimation, however, the *score orthogonality* condition is just a necessary condition for optimality but not a sufficient one. Next, we will present an example to demonstrate that if an estimator satisfies the score orthogonality condition, it can be a local minimum or even a local maximum of $H(e)$ in a certain direction. Before proceeding, we give a definition.

Definition 3.1 Given an estimator $g \in G$, g is said to be a local minimum (or maximum) of $H(e)$ in the direction of $\varphi \in G$, if and only if $\exists \varepsilon > 0$, such that $\forall\, \gamma \in \mathbb{R}$, $|\gamma| \leq \varepsilon$, we have

$$H(e|g) \overset{(\geq)}{\leq} H(e|g + \gamma\varphi) \tag{3.55}$$

Example 3.1 [92] Suppose the joint PDF of X and Y is the mix-Gaussian density $(0 \leq |\rho| < 1)$:

$$p(x,y) = \frac{1}{4\pi\sqrt{1-\rho^2}} \left\{ \begin{array}{l} \exp\left\{ \dfrac{-(y^2 - 2\rho y(x-\mu) + (x-\mu)^2)}{2(1-\rho^2)} \right\} \\[4mm] + \exp\left\{ \dfrac{-(y^2 + 2\rho y(x+\mu) + (x+\mu)^2)}{2(1-\rho^2)} \right\} \end{array} \right\} \tag{3.56}$$

The MMSE estimation of X based on Y can be computed as $g_{\text{MMSE}} = E(X|Y) = 0$. It is easy to check that the estimator g_{MMSE} satisfies the score orthogonality condition.

Now we examine whether the MMSE estimator g_{MMSE} is a local minimum or maximum of the error entropy $H(e)$ in a certain direction $\varphi \in G$. We focus here on the case where $\varphi(Y) = Y$ (linear function). In this case, we have

$$H(e|g_{\text{MMSE}}(Y) + \gamma\varphi(Y)) = H(e|\gamma Y) \tag{3.57}$$

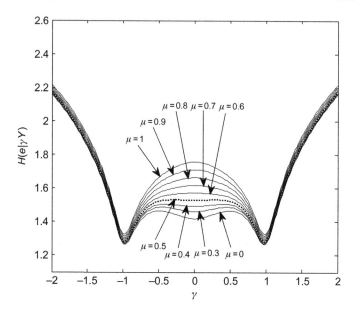

Figure 3.1 Error's entropy $H(e|\gamma Y)$ with respect to different γ and μ values.
Source: Adopted from [92].

For the case $\rho = 0.99$, the error's entropies $H(e|\gamma Y)$ with respect to different γ and μ values are shown in Figure 3.1, from which we see that the MMSE estimator ($\gamma = 0$) is a local minimum of the error entropy in the direction of $\varphi(Y) = Y$ for $\mu \leq 0.5$ and a local maximum for $\mu > 0.5$. In addition, for the case $\mu = 1.0$, the error's entropies $H(e|\gamma Y)$ with respect to different γ and ρ values are depicted in Figure 3.2. It can be seen from Figure 3.2 that when $\rho \leq 0.6$, g_{MMSE} is a global minimum of $H(e|\gamma Y)$; while when $\rho > 0.6$, it becomes a local maximum.

The local minima or local maxima can also be judged using the second-order derivative of error entropy with respect to γ. For instance, if $\mu = 1$, we can calculate the following second-order derivative using results of [167]:

$$\frac{\partial^2}{\partial \gamma^2} H(e|\gamma Y)\Big|_{\gamma=0} = (1 - \rho^2)E\left\{\left[\frac{\exp(2e)-1}{\exp(2e)+1}\right]^2\right\}$$

$$+ \rho^2 E\left\{\frac{[2(\exp(2e)-1)^2(e^2-1)\exp(2e)] - [\exp(4e)-1+4e\times\exp(2e)]^2}{[\exp(2e)+1]^4}\right\}$$

$$\approx 0.55 - 1.54\rho^2$$

$$(3.58)$$

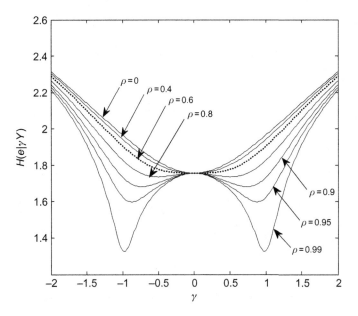

Figure 3.2 Error's entropy $H(e|\gamma Y)$ with respect to different γ and ρ values. *Source:* Adopted from [92].

And hence

$$
\begin{cases}
\frac{\partial^2}{\partial \gamma^2} H(e|\gamma Y)\big|_{\gamma=0} > 0 & \text{if } |\rho| < 0.6 \\
\frac{\partial^2}{\partial \gamma^2} H(e|\gamma Y)\big|_{\gamma=0} < 0 & \text{if } |\rho| > 0.6
\end{cases}
\tag{3.59}
$$

which implies that if $|\rho| < 0.6$, the MMSE estimator ($\gamma = 0$) will be a local minimum of the error entropy in the direction of $\varphi(Y) = Y$, whereas if $|\rho| > 0.6$, it becomes a local maximum.

As can be seen from Figure 3.2, if $\rho = 0.9$, the error entropy $H(e|\gamma Y)$ achieves its global minima at $\gamma \approx \pm 0.74$. Figure 3.3 depicts the error PDF for $\gamma = 0$ (MMSE estimator) and $\gamma = 0.74$ (linear MEE estimator), where $\mu = 1, \rho = 0.9$. We can see that the MEE solution is in this case not unique but it is much more concentrated (with higher peak) than the MMSE solution, which potentially gives an estimator with much smaller variance. We can also observe that the peaks of the MMSE and MEE error distributions occur at two different error locations, which means that the best parameter sets are different for each case.

3.3.1.2 Relationship to Conventional Bayes Risks

The loss function of MEE criterion is directly related to the error's PDF, which is much different from the loss functions in conventional Bayes risks, because the

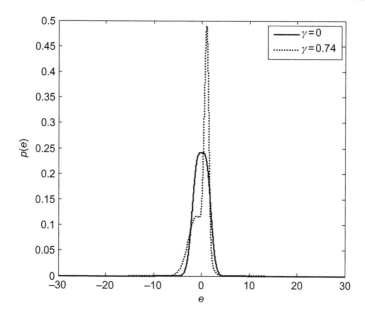

Figure 3.3 Error's PDF $p(e)$ for $\gamma = 0$ and $\gamma = 0.74$.
Source: Adopted from [92].

user does not need to select the risk functional. To some extent, the error entropy can be viewed as an "adaptive" Bayes risk, in which the loss function is varying with the error distribution, that is, different error distributions correspond to different loss functions. Figure 3.4 shows the loss functions (the lower subplots) of MEE corresponding to three different error PDFs. Notice that the third case provides a risk function that is nonconvex in the space of the errors. This is an unconventional risk function because the role of the weight function is to privilege one solution versus all others in the space of the errors.

There is an important relationship between the MEE criterion and the traditional MSE criterion. The following theorem shows that the MSE is equivalent to the error entropy plus the KL-divergence between the error PDF and any zero-mean Gaussian density.

Theorem 3.1 Let $G_\sigma(.)$ denote a Gaussian density, $G_\sigma(x) = (1/\sqrt{2\pi}\sigma)\exp(-x^2/2\sigma^2)$, where $\sigma > 0$. Then we have

$$\min_{g \in G} E(e^2) \Leftrightarrow \min_{g \in G}\{H(e) + D_{KL}(p_e \| G_\sigma)\} \tag{3.60}$$

Proof: Since $G_\sigma(x) = (1/\sqrt{2\pi}\sigma)\exp(-x^2/2\sigma^2)$, we have

$$x^2 = -2\sigma^2\left\{\log(G_\sigma(x)) + \log(\sqrt{2\pi}\sigma)\right\}$$

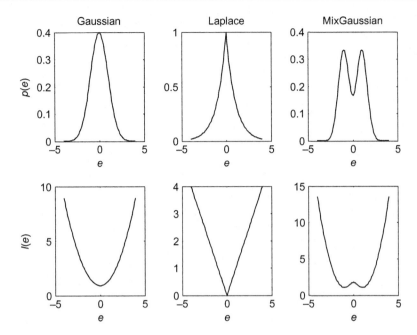

Figure 3.4 The loss functions of MEE corresponding to three different error PDFs.

And hence

$$
\begin{aligned}
E(e^2) &= \int x^2 p_e(x) \mathrm{d}x \\
&= 2\sigma^2 \left\{ -\int (\log G_\sigma(x)) p_e(x) \mathrm{d}x - \log\left(\sqrt{2\pi}\sigma\right) \right\} \\
&= 2\sigma^2 \left\{ -\int p_e(x)\log p_e(x) \mathrm{d}x + \int p_e(x)\left(\log \frac{p_e(x)}{G_\sigma(x)}\right)\mathrm{d}x - \log\left(\sqrt{2\pi}\sigma\right) \right\} \\
&= 2\sigma^2 \left\{ H(e) + D_{\mathrm{KL}}(p_e \| G_\sigma) - \log\left(\sqrt{2\pi}\sigma\right) \right\}
\end{aligned}
$$

$$(3.61)$$

It follows easily that $\min\limits_{g \in G} E(e^2) \Leftrightarrow \min\limits_{g \in G} \{H(e) + D_{\mathrm{KL}}(p_e \| G_\sigma)\}$.

Remark: The above theorem suggests that the minimization of MSE minimizes both the error entropy $H(e)$ and the KL-divergence $D_{\mathrm{KL}}(p_e \| G_\sigma)$. Then the MMSE estimation will decrease the error entropy, and at the same time, make the error distribution close to zero-mean Gaussian distribution. In nonlinear and non-Gaussian estimating systems, the desirable error distribution can be far from Gaussian, while the MSE criterion still makes the error distribution close to zero-mean Gaussian distribution, which may lead to poor estimation performance. Thus, Theorem 3.1 also gives an explanation on why the performance of MMSE estimation may be not good for non-Gaussian samples.

Suppose that the error is zero-mean Gaussian distributed. Then we have $D_{\mathrm{KL}}(p_e\|G_\sigma) = 0$, where $\sigma^2 = E[e^2]$. In this case, the MSE criterion is equivalent to the MEE criterion.

From the derivation of Theorem 3.1, let $\sigma = \sqrt{2}/2$, we have

$$E(e^2) = H(e) + D_{\mathrm{KL}}(p_e\|G_{\sqrt{2}/2}) - \log(\sqrt{\pi}) \tag{3.62}$$

Since $D_{\mathrm{KL}}(p_e\|G_{\sqrt{2}/2}) \geq 0$, then

$$H(e) \leq E(e^2) + \log(\sqrt{\pi}) \tag{3.63}$$

Inequality (3.63) suggests that, minimizing the MSE is equivalent to minimizing an upper bound of error entropy.

There exists a similar relationship between MEE criterion and a large family of Bayes risks [168]. To prove this fact, we need a lemma.

Lemma 3.1 Any Bayes risk $E[l(e)]$ corresponds to a PDF[4] as follows:

$$q_l(x) = \exp[-\gamma_0 - \gamma_1 l(x)] \tag{3.64}$$

where γ_0 and γ_1 satisfy

$$\begin{cases} \exp(\gamma_0) = \int_{\mathbb{R}} \exp[-\gamma_1 l(x)]\mathrm{d}x \\ E[l(e)]\exp(\gamma_0) = \int_{\mathbb{R}} l(x)\exp[-\gamma_1 l(x)]\mathrm{d}x \end{cases} \tag{3.65}$$

Theorem 3.2 For any Bayes risk $E[l(e)]$, if the loss function $l(e)$ satisfies $\lim_{|e|\to+\infty} l(e) = +\infty$, then

$$\min_{g\in G} E[l(e)] \Leftrightarrow \min_{g\in G}\{H(e) + D_{\mathrm{KL}}(p_e\|q_l)\} \tag{3.66}$$

where q_l is the PDF given in (3.64).

Proof: First, we show that in the PDF $q_l(x) = \exp[-\gamma_0 - \gamma_1 l(x)]$, γ_1 is a positive number. Since $q_l(.)$ satisfies $q_l(x) \geq 0$ and $\int_{\mathbb{R}} q_l(x)\mathrm{d}x = 1$, we have $\lim_{|x|\to+\infty} q_l(x) = 0$. Then

---

[4] Here, $q_l(.)$ is actually the maximum entropy density that satisfies the constraint condition $\int_{\mathbb{R}} q(x)l(x)\mathrm{d}x = E[l(e)]$.

$$\lim_{|x|\to +\infty} q_l(x) = 0$$

$$\Rightarrow \lim_{|x|\to\infty} [\log q_l(x)] = -\infty$$

$$\Rightarrow \lim_{|x|\to +\infty} [-\log q_l(x)] = +\infty$$

$$\Rightarrow \lim_{|x|\to +\infty} [\gamma_0 + \gamma_1 l(x)] = +\infty \tag{3.67}$$

$$\Rightarrow \gamma_1 \lim_{|x|\to +\infty} l(x) = +\infty$$

$$\overset{(a)}{\Rightarrow} \gamma_1 > 0$$

where (a) follows from $\lim_{|x|\to +\infty} l(x) = +\infty$. Therefore,

$$\min_{g\in G}\{H(e) + D_{KL}(p_e\|q_l)\}$$

$$\Leftrightarrow \min_{g\in G}\left\{-\int p_e(x)\log p_e(x)dx + \int p_e(x)\log\left(\frac{p_e(x)}{q_l(x)}\right)dx\right\}$$

$$\Leftrightarrow \min_{g\in G}\left\{\int p_e(x)[-\log q_l(x)]dx\right\} \tag{3.68}$$

$$\Leftrightarrow \min_{g\in G}\left\{\int p_e(x)[\gamma_0 + \gamma_1 l(x)]dx\right\}$$

$$\Leftrightarrow \min_{g\in G}\{\gamma_0 + \gamma_1 E[l(e)]\}$$

$$\overset{\gamma_1 > 0}{\Leftrightarrow} \min_{g\in G}\{E[l(e)]\}$$

which completes the proof.

Remark: The condition $\lim_{|e|\to +\infty} l(e) = +\infty$ in the theorem is not very restrictive, because for most Bayes risks, the loss function increases rapidly when $|e|$ goes to infinity. But for instance, it does not apply to the maximum correntropy (MC) criterion studied next.

3.3.2 MC Estimation

Correntropy is a novel measure of similarity between two random variables [64]. Let X and Y be two random variables with the same dimensions, the correntropy is

$$V(X, Y) = E[\kappa(X, Y)] \tag{3.69}$$

where $\kappa(.,.)$ is a translation invariant Mercer kernel. The most popular kernel used in correntropy is the Gaussian kernel:

$$\kappa_\sigma(x, y) = \frac{1}{\sqrt{2\pi}\sigma} \exp(- \|x - y\|^2/2\sigma^2) \qquad (3.70)$$

where $\sigma > 0$ denotes the kernel size (kernel width). Gaussian kernel $\kappa_\sigma(x, y)$ is a translation invariant kernel that is a function of $x - y$, so it can be rewritten as $\kappa_\sigma(x - y)$.

Compared with other similarity measures, such as the mean square error, correntropy (with Gaussian kernel) has some nice properties: (i) it is always bounded $(0 < V(X, Y) \le 1/\sqrt{2\pi}\sigma)$; (ii) it contains all even-order moments of the difference variable for the Gaussian kernel (using a series expansion); (iii) the weights of higher order moments are controlled by kernel size; and (iv) it is a local similarity measure, and is very robust to outliers.

The correntropy function can also be applied to Bayes estimation [169]. Let X be an unknown parameter to be estimated and Y be the observation. We assume, for simplification, that X is a scalar random variable (extension to the vector case is straightforward), $X \in \mathbb{R}$, and Y is a random vector taking values in $\mathbb{R}^m$. The MC estimation of X based on Y is to find a measurable function $g:\mathbb{R}^m \to \mathbb{R}$ such that the correntropy between X and $\hat{X} = g(Y)$ is maximized, i.e.,

$$g_{MC} = \arg \max_{g \in G} E[\kappa(X, g(Y))] \qquad (3.71)$$

With any translation invariant kernel such as the Gaussian kernel $\kappa_\sigma(x - y)$, the MC estimator will be

$$g_{MC} = \arg \max_{g \in G} E[\kappa_\sigma(e)] \qquad (3.72)$$

where $e = X - g(Y)$ is the estimation error. If $\forall y \in \mathbb{R}^m$, X has posterior PDF $p(x|y)$, then the estimation error has PDF

$$p_e(x) = \int_{\mathbb{R}^m} p(x + g(y)|y)dF(y) \qquad (3.73)$$

where $F(y)$ denotes the distribution function of Y. In this case, we have

$$\begin{aligned} g_{MC} &= \arg \max_{g \in G} E[\kappa_\sigma(e)] \\ &= \arg\max_{g \in G} \int_{-\infty}^{\infty} \kappa_\sigma(x) \int_{\mathbb{R}^m} p(x + g(y)|y)dF(y)dx \end{aligned} \qquad (3.74)$$

The following theorem shows that the MC estimation is a smoothed MAP estimation.

Theorem 3.3 The MC estimator (3.74) can be expressed as

$$g_{MC}(y) = \arg\max_{x \in R} \rho(x|y, \sigma), \quad \forall\, y \in R^m \tag{3.75}$$

where $\rho(x|y, \sigma) = \kappa_\sigma(x) * p(x|y)$ ("$*$" denotes the convolution operator with respect to x).

Proof: One can derive

$$
\begin{aligned}
V(e) &= \int_{-\infty}^{\infty} \kappa_\sigma(x) \int_{R^m} p(x + g(y)|y)\mathrm{d}F(y)\mathrm{d}x \\
&= \int_{R^m} \left\{ \int_{-\infty}^{\infty} \kappa_\sigma(x)p(x + g(y)|y)\mathrm{d}x \right\}\mathrm{d}F(y) \\
&= \int_{R^m} \left\{ \int_{-\infty}^{\infty} \kappa_\sigma(x' - g(y))p(x'|y)\mathrm{d}x' \right\}\mathrm{d}F(y) \\
&\overset{(a)}{=} \int_{R^m} \left\{ \int_{-\infty}^{\infty} \kappa_\sigma(g(y) - x')p(x'|y)\mathrm{d}x' \right\}\mathrm{d}F(y) \\
&= \int_{R^m} \{(\kappa_\sigma(.) * p(.|y))(g(y))\}\mathrm{d}F(y) \\
&= \int_{R^m} \rho(g(y)|y, \sigma)\mathrm{d}F(y)
\end{aligned}
\tag{3.76}
$$

where $x' = x + g(y)$ and (a) comes from the symmetry of $\kappa_\sigma(.)$. It follows easily that

$$
\begin{aligned}
g_{MC} &= \arg\max_{g \in G} \int_{R^m} \rho(g(y)|y, \sigma)\mathrm{d}F(y) \\
\Rightarrow g_{MC}(y) &= \arg\max_{x \in R} \rho(x|y, \sigma), \quad \forall\, y \in R^m
\end{aligned}
\tag{3.77}
$$

This completes the proof.

Remark: The function $\rho(x|y, \sigma)$ can be viewed as a smoothed version (through convolution) of the posterior PDF $p(x|y)$. Thus according to Theorem 3.3, the MC estimation is in essence a smoothed MAP estimation, which is the mode of the smoothed posterior distribution. The kernel size σ plays an important role in the smoothing process by controlling the degree of smoothness. When $\sigma \to 0+$, the Gaussian kernel will approach the Dirac delta function, and the function $\rho(x|y, \sigma)$ will reduce to the original posterior PDF. In this case, the MC estimation is identical to the MAP estimation. On the other hand, when $\sigma \to \infty$, the second-order moment will dominate the correntropy, and the MC estimation will be equivalent to the MMSE estimation [137].

In the literature of global optimization, the convolution smoothing method [170–172] has been proven very effective in searching the global minima (or maxima). Usually, one can use the convolution of a nonconvex cost function with a suitable smooth function to eliminate local optima, and gradually decrease the degree of smoothness to achieve global optimization. Therefore, we believe that by properly annealing the kernel size, the MC estimation can be used to obtain the global maxima of MAP estimation.

The optimal solution of MC estimation is the mode of the smoothed posterior distribution, which is, obviously, not necessarily unique. The next theorem, however, shows that when kernel size is larger than a certain value, the MC estimation will have a unique optimal solution that lies in a strictly concave region of the smoothed PDF (see [169] for the proof).

Theorem 3.4 [169] Assume that function $f_n(y) = \int_{-n}^{n} p(x|y)dx$ converges uniformly to 1 as $n \to \infty$. Then there exists an interval $[-M, M]$, $M > 0$, such that when kernel size σ is larger than a certain value, the smoothed PDF $\rho(x|y, \sigma)$ will be strictly concave in $[-M, M]$, and has a unique global maximum lying in this interval for any $y \in \mathbb{R}^m$.

Remark: Theorem 3.4 suggests that in MC estimation, one can use a larger kernel size to eliminate local optima by constructing a concave (in a certain interval) cost function with a unique global optimal solution. This result also shows that the initial condition for convolution smoothing should be chosen in this range. The only other parameter in the method that the user has to select is the annealing rate.

In the following, a simple example is presented to illustrate how the kernel size affects the solution of MC estimation. Suppose the joint PDF of X and Y $(X, Y \in \mathbb{R})$ is the mixture density $(0 \le \lambda \le 1)$ [169]:

$$p_{XY}(x, y) = (1 - \lambda)p_1(x, y) + \lambda p_2(x, y) \tag{3.78}$$

where p_1 denotes the "clean" PDF and p_2 denotes the contamination part corresponding to "bad" data or outliers. Let $\lambda = 0.03$, and assume that p_1 and p_2 are both jointly Gaussian:

$$p_1(x, y) = \frac{1}{\sqrt{3}\pi} \exp\left(-\frac{x^2 - xy + y^2}{1.5}\right) \tag{3.79}$$

$$p_2(x, y) = \frac{50}{\pi} \exp\left(-50\left[(x-3)^2 + y^2\right]\right) \tag{3.80}$$

For the case $y = 0.1$, the smoothed posterior PDFs with different kernel sizes are shown in Figure 3.5, from which we observe: (i) when kernel size is small, the smoothed PDFs are nonconcave within the dominant region (say the interval $[-3, 4]$), and there may exist local optima or even nonunique optimal solutions; (ii)

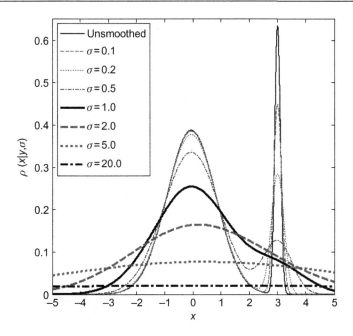

Figure 3.5 Smoothed conditional PDFs given $Y = 0.1$.
Source: Adopted from [169]).

Table 3.2 Estimates of X Given $Y = 0.1$

MAP estimation	3.0	
MMSE estimation	0.4338	
MC estimation	$\sigma = 0.1$	3.0
	$\sigma = 0.2$	-0.05
	$\sigma = 0.5$	-0.05
	$\sigma = 1.0$	-0.0360
	$\sigma = 2.0$	0.2090
	$\sigma = 5.0$	0.3990
	$\sigma = 20$	0.4320

(adopted from [169])

when kernel size is larger, the smoothed PDFs become concave within the dominant region, and there is a unique optimal solution. Several estimates of X given $Y = 0.1$ are listed in Table 3.2. It is evident that when the kernel size is very small, the MC estimate is the same as the MAP estimate; while when the kernel size is very large, the MC estimate is close to the MMSE estimate. In particular, for some kernel sizes (say $\sigma = 0.2$ or 0.5), the MC estimate of X equals -0.05, which is exactly the MMSE (or MAP) estimate of X based on the "clean" distribution p_1. This result confirms the fact that the MC estimation is much more robust (with respect to outliers) than both MMSE and MAP estimations.

3.4 Information Criteria for Model Selection

Information theoretic approaches have also been used to solve the model selection problem. Consider the problem of estimating the parameter θ of a family of models

$$\mathcal{M} = \{p(x, \theta)|\theta \in \Theta_k \subset \mathbb{R}^k\} \tag{3.81}$$

where the parameter space dimension k is also unknown. This problem is actually a model structure selection problem, and the value of k is the structure parameter. There are many approaches to select the space dimension k. Here, we only discuss several *information criteria* for selecting the most parsimonious correct model, where the name indicates that they are closely related to or can be derived from information theory.

1. *Akaike's information criterion*

The *Akaike's information criterion* (AIC) was first developed by Akaike [6,7]. AIC is a measure of the relative goodness of fit of a statistical model. It describes the tradeoff between bias and variance (or between accuracy and complexity) in model construction. In the general case, AIC is defined as

$$\text{AIC} = -2 \log L_{\max} + 2k \tag{3.82}$$

where $L_{\max}$ is the maximized value of the likelihood function for the estimated model. To apply AIC in practice, we start with a set of candidate models, and find the models' corresponding AIC values, and then select the model with the minimum AIC value. Since AIC includes a penalty term $2k$, it can effectively avoid overfitting. However, the penalty is constant regardless of the number of samples used in the fitting process.

The AIC criterion can be derived from the KL-divergence minimization principle or the equivalent relative entropy maximization principle (see Appendix G for the derivation).

2. *Bayesian Information Criterion*

The *Bayesian information criterion* (BIC), also known as the Schwarz criterion, was independently developed by Akaike and by Schwarz in 1978, using Bayesian formalism. Akaike's version of BIC was often referred to as the ABIC (for "a BIC") or more casually, as Akaike's Bayesian Information Criterion. BIC is based, in part, on the likelihood function, and is closely related to AIC criterion. The formula for the BIC is

$$\text{BIC} = -2 \log L_{\max} + k \log n \tag{3.83}$$

where n denotes the number of the observed data (i.e., sample size). The BIC criterion has a form very similar to AIC, and as one can see, the penalty term in BIC is in general larger than in AIC, which means that generally it will provide smaller model sizes.

3. *Minimum Description Length Criterion*

The *minimum description length* (MDL) principle was introduced by Rissanen [9]. It is an important principle in information and learning theories. The fundamental idea behind the MDL principle is that any regularity in a given set of data can be used to compress the data, that is, to describe it using fewer symbols than needed to describe the data literally. According to MDL, the best model for a given set of data is the one that leads to the best compression of the data. Because data compression is formally equivalent to a form of probabilistic prediction, MDL methods can be interpreted as searching for a model with good predictive performance on unseen data. The ideal MDL approach requires the estimation of the Kolmogorov complexity, which is noncomputable in general. However, there are nonideal, practical versions of MDL.

From a coding perspective, assume that both sender and receiver know which member $p(x, \theta)$ of the parametric family $\mathscr{M}$ generated a data string $x^n = \{x_1, x_2, \ldots, x_n\}$. Then from a straightforward generalization of Shannon's Source Coding Theorem to continuous random variables, it follows that the best description length of x^n (in an average sense) is simply $-\log p(x^n, \theta)$, because on average the code length achieves the entropy lower bound $- \int p(x^n, \theta)\log p(x^n, \theta)\mathrm{d}x^n$. Clearly, minimizing $-\log p(x^n, \theta)$ is equivalent to maximizing $p(x^n, \theta)$. Thus the MDL coincides with the ML in parametric estimation problems. In addition, we have to transmit θ, because the receiver did not know its value in advance. Adding in this cost, we arrive at a code length for the data string x^n:

$$-\log p(x^n, \hat{\theta}_{\mathrm{ML}}) + l(\hat{\theta}_{\mathrm{ML}}) \tag{3.84}$$

where $l(\hat{\theta}_{\mathrm{ML}})$ denotes the number of bits for transmitting $\hat{\theta}_{\mathrm{ML}}$. If we assume that the machine precision is $1/\sqrt{n}$ for each component of $\hat{\theta}_{\mathrm{ML}}$ and $\hat{\theta}_{\mathrm{ML}}$ is transmitted with a uniform encoder, then the term $l(\theta)$ is expressed as

$$l(\theta) = \frac{k}{2}\log n \tag{3.85}$$

In this case, the MDL takes the form of BIC. An alternative expression of $l(\theta)$ is

$$l(\theta) = \sum_{j=1}^{k} \log\frac{\gamma}{\delta_j} \tag{3.86}$$

where γ is a constant related to the number of bits in the exponent of the floating point representation of θ_j and δ_j is the optimal precision of θ_j.

Appendix E: EM Algorithm

An EM algorithm is an iterative method for finding the ML estimate of parameters in statistical models, where the model depends on unobserved latent variables. The EM iteration alternates between performing an expectation (E) step, which calculates the expectation of the log-likelihood evaluated using the current estimate for the parameters, and a maximization (M) step, which computes parameters maximizing the expected log-likelihood found on the E step. These parameter estimates are then used to determine the distribution of the latent variables in the next E step.

Let y be the observed data, z be the unobserved data, and θ be a vector of unknown parameters. Further, let $L(\theta|y, z)$ be the likelihood function, $L(\theta|y)$ be the marginal likelihood function of the observed data, and $p(z|y, \theta^{(k)})$ be the conditional density of z given y under the current estimate of the parameters $\theta^{(k)}$. The EM algorithm seeks to find the ML estimate of the marginal likelihood by iteratively applying the following two steps

1. Expectation step (E step): Calculate the expected value of the log-likelihood function, with respect to the conditional distribution of z given y under the current estimate $\theta^{(k)}$:

$$
\begin{aligned}
Q(\theta|y, \theta^{(k)}) &= E_{z|y, \theta^{(k)}}[\log L(\theta|y, z)|y, \theta^{(k)}] \\
&= \int \log L(\theta|y, z)p(z|y, \theta^{(k)})dz
\end{aligned}
\tag{E.1}
$$

2. Maximization step (M step): Find the next estimate $\theta^{(k+1)}$ of the parameters by maximizing this quantity:

$$
Q(\theta^{(k+1)}|y, \theta^{(k)}) = \max_{\theta} Q(\theta|y, \theta^{(k)})
\tag{E.2}
$$

The iteration $\theta^{(k)} \to \theta^{(k+1)}$ continues until $\|\theta^{(k+1)} - \theta^{(k)}\|$ is sufficiently small.

Appendix F: Minimum MSE Estimation

The MMSE estimate $\hat{\theta}$ of θ is obtained by minimizing the following cost:

$$
\begin{aligned}
R(\theta, \hat{\theta}) &= \int \int (\hat{\theta} - \theta)^2 p(\theta, x)d\theta \, dx \\
&= \int \left(\int (\hat{\theta} - \theta)^2 p(\theta|x)d\theta \right) p(x)dx
\end{aligned}
\tag{F.1}
$$

Since $\forall x, p(x) \geq 0$, one only needs to minimize $\int (\hat{\theta} - \theta)^2 p(\theta|x)d\theta$. Let

$$
\frac{\partial}{\partial \hat{\theta}} \int (\hat{\theta} - \theta)^2 p(\theta|x)d\theta = 2 \int (\hat{\theta} - \theta)p(\theta|x)d\theta = 0
\tag{F.2}
$$

Then we get $\hat{\theta} = \int \theta p(\theta|x)d\theta$.

Appendix G: Derivation of AIC Criterion

The information theoretic KL-divergence plays a crucial role in the derivation of AIC. Suppose that the data are generated from some distribution f. We consider two candidate models (distributions) to represent f: g_1 and g_2. If we know f, then we could evaluate the information lost from using g_1 to represent f by calculating the KL-divergence, $D_{KL}(f \| g_1)$; similarly, the information lost from using g_2 to represent f would be found by calculating $D_{KL}(f \| g_2)$. We would then choose the candidate model minimizing the information loss. If f is unknown, we can estimate, via AIC, how much more (or less) information is lost by g_1 than by g_2. The estimate is, certainly, only valid asymptotically.

Given a family of density models $\{p_\theta(y)|\theta \in \Theta_k \subset \mathbb{R}^k\}$, where Θ_k denotes a k-dimensional parameter space, $k = 1, 2, \ldots, K$, and a sequence of independent and identically distributed observations $\{y_1, y_2, \ldots, y_n\}$, the AIC can be expressed as

$$\text{AIC} = -2 \log L(\hat{\theta}_{\text{ML}}) + 2k \tag{G.1}$$

where $\hat{\theta}_{\text{ML}}$ is the ML estimate of $\theta = [\theta_1, \theta_2, \ldots, \theta_k]$:

$$\hat{\theta}_{\text{ML}} = \arg \max_{\theta \in \Theta_k} L(\theta) = \arg \max_{\theta \in \Theta_k} \prod_{i=1}^{n} p_\theta(y_i) \tag{G.2}$$

Let θ_0 be the unknown true parameter vector. Then

$$D_{\text{KL}}(\theta_0 \| \hat{\theta}_{\text{ML}}) = D_{\text{KL}}(p_{\theta_0} \| p_{\hat{\theta}_{\text{ML}}}) = E\{\log p_{\theta_0}(y)\} - E\{\log p_{\hat{\theta}_{\text{ML}}}(y)\} \tag{G.3}$$

where

$$\begin{cases} E\{\log p_{\theta_0}(y)\} = \int_{-\infty}^{\infty} p_{\theta_0}(y) \log p_{\theta_0}(y) dy \\ E\{\log p_{\hat{\theta}_{\text{ML}}}(y)\} = \int_{-\infty}^{\infty} p_{\theta_0}(y) \log p_{\hat{\theta}_{\text{ML}}}(y) dy \end{cases} \tag{G.4}$$

Taking the first term in a Taylor expansion of $E\{\log p_{\hat{\theta}_{\text{ML}}}(y)\}$, we obtain

$$\{\log p_{\hat{\theta}_{\text{ML}}}(y)\} \approx E\{\log p_{\theta_0}(y)\} - \frac{1}{2}(\hat{\theta}_{\text{ML}} - \theta_0)^T J_F(\theta_0)(\hat{\theta}_{\text{ML}} - \theta_0) \tag{G.5}$$

where $J_F(\theta_0)$ is the $k \times k$ Fisher information matrix. Then we have

$$D_{\text{KL}}(\theta_0 \| \hat{\theta}_{\text{ML}}) \approx \frac{1}{2}(\hat{\theta}_{\text{ML}} - \theta_0)^T J_F(\theta_0)(\hat{\theta}_{\text{ML}} - \theta_0) \tag{G.6}$$

Suppose $J_F(\theta_0)$ can be decomposed into

$$J_F(\theta_0) = J^T J \tag{G.7}$$

where J is some nonsingular matrix. We can derive

$$2n D_{\text{KL}}(\theta_0 \| \hat{\theta}_{\text{ML}}) \approx [\sqrt{n} J(\hat{\theta}_{\text{ML}} - \theta_0)]^T [\sqrt{n} J(\hat{\theta}_{\text{ML}} - \theta_0)] \tag{G.8}$$

According to the statistical properties of the ML estimator, when sample number n is large enough, we have

$$\sqrt{n}(\hat{\theta}_{\text{ML}} - \theta_0) \sim \mathcal{N}(0, J_F^{-1}(\theta_0)) \tag{G.9}$$

Combining (G.9) and (G.7) yields

$$\sqrt{n}J(\hat{\theta}_{ML} - \theta_0) \sim \mathcal{N}(0, I) \tag{G.10}$$

where I is the $k \times k$ identity matrix. From (G.8) and (G.10), we obtain

$$2nD_{KL}(\theta_0 || \hat{\theta}_{ML}) \sim \chi^2(k) \tag{G.11}$$

That is, $2nD_{KL}(\theta_0 || \hat{\theta}_{ML})$ is Chi-Squared distributed with k degree of freedom. This implies

$$E\{2nD_{KL}(\theta_0 || \hat{\theta}_{ML})\} = k \tag{G.12}$$

It follows that

$$2nE\{\log p_{\theta_0}(y)\} - 2nE\{\log p_{\hat{\theta}_{ML}}(y)\} \xrightarrow[n \to \infty]{a.s.} k \tag{G.13}$$

And hence

$$2nE\{\log p_{\hat{\theta}_{ML}}(y)\} \xrightarrow[n \to \infty]{a.s.} 2\log L(\theta_0) - k \tag{G.14}$$

It has been proved in [173] that

$$2[\log L(\hat{\theta}_{ML}) - \log L(\theta_0)] \sim \chi^2(k) \tag{G.15}$$

Therefore

$$2\log L(\theta_0) \xrightarrow[n \to \infty]{a.s.} 2\log L(\hat{\theta}_{ML}) - k \tag{G.16}$$

Combining (G.16) and (G.14), we have

$$2nE\{\log p_{\hat{\theta}_{ML}}(y)\} \xrightarrow[n \to \infty]{a.s.} 2\log L(\hat{\theta}_{ML}) - 2k \tag{G.17}$$

To minimize the KL-divergence $D_{KL}(\theta_0 || \hat{\theta}_{ML})$, one need to maximize $E\{\log p_{\hat{\theta}_{ML}}(y)\}$, or equivalently, to minimize (in an asymptotical sense) the following objective function

$$-2\log L(\hat{\theta}_{ML}) + 2k \tag{G.18}$$

This is exactly the AIC criterion.

4 System Identification Under Minimum Error Entropy Criteria

In previous chapter, we give an overview of information theoretic parameter estimation. These estimation methods are, however, devoted to cases where a large amount of statistical information on the unknown parameter is assumed to be available. For example, the minimum divergence estimation needs to know the likelihood function of the parameter. Also, in Bayes estimation with minimum error entropy (MEE) criterion, the joint distribution of unknown parameter and observation is assumed to be known. In this and later chapters, we will further investigate information theoretic system identification. Our focus is mainly on system parameter estimation (identification) where no statistical information on parameters exists (i.e., only data samples are available). To develop the identification algorithms under information theoretic criteria, one should evaluate the related information measures. This requires the knowledge of the data distributions, which are, in general, unknown to us. To address this issue, we can use the estimated (empirical) information measures as the identification criteria.

4.1 Brief Sketch of System Parameter Identification

System identification involves fitting the experimental input−output data (training data) into empirical model. In general, system identification includes the following key steps:

- Experiment design: To obtain good experimental data. Usually, the input signals should be designed such that it provides enough process excitation.
- Selection of model structure: To choose a suitable model structure based on the training data or prior knowledge.
- Selection of the criterion: To choose a suitable criterion (cost) function that reflects how well the model fits the experimental data.
- Parameter identification: To obtain the model parameters[1] by optimizing (minimizing or maximizing) the above criterion function.
- Model validation: To test the model so as to reveal any inadequacies.

In this book, we focus mainly on the parameter identification part. Figure 4.1 shows a general scheme of discrete-time system identification, where x_k and y_k

---

[1] In the case of black-box identification, the model parameters are basically viewed as vehicles for adjusting the fit to data and do not reflect any physical consideration in the system.

System Parameter Identification. DOI: http://dx.doi.org/10.1016/B978-0-12-404574-3.00004-X
© 2013 Tsinghua University Press Ltd. Published by Elsevier Inc. All rights reserved.

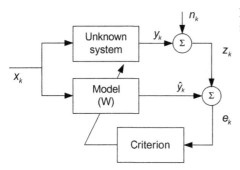

Figure 4.1 A general scheme of system identification.

denote the system input and output (clean output) at time k, n_k is an additive noise that accounts for the system uncertainty or measurement error, and z_k is the measured output. Further, $\hat{y}_k$ is the model output and e_k denotes the identification error, which is defined as the difference between the measured output and the model output, i.e., $e_k = z_k - \hat{y}_k$. The goal of parameter identification is then to search the model parameter vector (or weight vector) so as to minimize (or maximize) a certain criterion function (usually the model structure is predefined).

The implementation of system parameter identification involves model structure, criterion function, and parameter search (identification) algorithm. In the following, we will briefly discuss these three aspects.

4.1.1 Model Structure

Generally speaking, the model structure is a parameterized mapping from inputs[2] to outputs. There are various mathematical descriptions of system model (linear or nonlinear, static or dynamic, deterministic or stochastic, etc.). Many of them can be expressed as the following *linear-in-parameter* model:

$$\begin{cases} z_k = \boldsymbol{h}_k^T W + e_k \\ \hat{y}_k = \boldsymbol{h}_k^T W \end{cases} \tag{4.1}$$

where $\boldsymbol{h}_k$ denotes the regression input vector and W denotes the weight vector (i.e., parameter vector). The simplest linear-in-parameter model is the adaptive linear neuron (ADALINE). Let the input be an m-dimensional vector $X_k = [x_{1,k}, \ldots, x_{m,k}]^T$. The output of ADALINE model will be

$$\hat{y}_k = \sum_{i=1}^{m} w_i x_{i,k} + w_0 \tag{4.2}$$

---

[2] For a dynamic system, the input vector may contain past inputs and outputs.

where w_0 is a bias (some constant). In this case, we have

$$\begin{cases} \boldsymbol{h}_k = [1, x_{1,k}, \ldots, x_{m,k}]^T \\ W = [w_0, w_1, \ldots, w_m]^T \end{cases} \tag{4.3}$$

If the bias w_0 is zero, and the input vector is $X_k = [x_k, x_{k-1}, \ldots, x_{k-m+1}]^T$, which is formed by feeding the input signal to a tapped delay line, then the ADALINE becomes a finite impulse response (FIR) filter.

The ARX (autoregressive with external input) dynamic model is another important linear-in-parameter model:

$$z_k + a_1 z_{k-1} + \cdots + a_{n_a} z_{k-n_a} = b_1 x_{k-1} + \cdots + b_{n_b} x_{k-n_b} + e_k \tag{4.4}$$

One can write Eq. (4.4) as $z_k = \boldsymbol{h}_k^T W + e_k$ if let

$$\begin{cases} \boldsymbol{h}_k = [-z_{k-1}, \ldots, -z_{k-n_a}, x_{k-1}, \ldots, x_{k-n_b}]^T \\ W = [a_1, \ldots, a_{n_a}, b_1, \ldots, b_{n_b}]^T \end{cases} \tag{4.5}$$

The linear-in-parameter model also includes many nonlinear models as special cases. For example, the n-order polynomial model can be expressed as

$$\hat{y}_k = a_0 + \sum_{i=1}^{n} a_i x_k^i = \boldsymbol{h}_k^T W \tag{4.6}$$

where

$$\begin{cases} \boldsymbol{h}_k = [1, x_k, \ldots, x_k^n]^T \\ W = [a_0, a_1, \ldots, a_n]^T \end{cases} \tag{4.7}$$

Other examples include: the discrete-time Volterra series with finite memory and order, Hammerstein model, radial basis function (RBF) neural networks with fixed centers, and so on.

In most cases, the system model is a *nonlinear-in-parameter* model, whose output is not linearly related to the parameters. A typical example of nonlinear-in-parameter model is the multilayer perceptron (MLP) [53]. The MLP, with one hidden layer, can be generally expressed as follows:

$$\hat{y}_k = W_2^T \phi(W_1^T X_k + b_1) + b_2 \tag{4.8}$$

where X_k is the $m \times 1$ input vector, W_1 is the $m \times n$ weight matrix connecting the input layer with the hidden layer, $\phi(.)$ is the activation function (usually a sigmoid function), b_1 is the $n \times 1$ bias vector for the hidden neurons, W_2 is the $n \times 1$ weight vector connecting the hidden layer to the output neuron, and b_2 is the bias for the output neuron.

The model can also be created in kernel space. In kernel machine learning (e.g. support vector machine, SVM), one often uses a reproducing kernel Hilbert space (RKHS) $\mathscr{H}_k$ associated with a Mercer kernel $\kappa: \mathscr{X} \times \mathscr{X} \rightarrow \mathbb{R}$ as the hypothesis space [161, 162]. According to Moore-Aronszajn theorem [174, 175], every Mercer kernel κ induces a unique function space $\mathscr{H}_\kappa$, namely the RKHS, whose reproducing kernel is κ, satisfying: 1) $\forall x \in \mathscr{X}$, the function $\kappa(x,.) \in \mathscr{H}_\kappa$, and 2) $\forall x \in \mathscr{X}$, and for every $f \in \mathscr{H}_\kappa$, $\langle f, \kappa(x,.) \rangle_{\mathscr{H}_\kappa} = f(x)$, where $\langle .,. \rangle_{\mathscr{H}_\kappa}$ denotes the inner product in $\mathscr{H}_\kappa$. If Mercer kernel κ is strictly positive-definite, the induced RKHS $\mathscr{H}_\kappa$ will be universal (dense in the space of continuous functions over $\mathscr{X}$). Assuming the input signal $x_k \in \mathscr{X}$, the model in RKHS $\mathscr{H}_\kappa$ can be expressed as

$$\hat{y}_k = f(x_k) = \langle f, \kappa(x_k,.) \rangle_{\mathscr{H}_\kappa} \qquad (4.9)$$

where $f \in \mathscr{H}_\kappa$ is the unknown input$-$output mapping that needs to be estimated. This model is a nonparametric function over input space $\mathscr{X}$. However, one can regard it as a "parameterized" model, where the parameter space is the RKHS $\mathscr{H}_\kappa$.

The model (4.9) can alternatively be expressed in a feature space (a vector space in which the training data are embedded). According to Mercer's theorem, any Mercer kernel κ induces a mapping φ from the input space $\mathscr{X}$ to a feature space $\mathbb{F}_\kappa$.[3] In the feature space, the inner products can be calculated using the kernel evaluation:

$$\varphi(x)^T \varphi(x') = \kappa(x, x') \qquad (4.10)$$

The feature space $\mathbb{F}_\kappa$ is isometric-isomorphic to the RKHS $\mathscr{H}_k$. This can be easily understood by identifying $\varphi(x) = \kappa(x,.)$ and $f = \Omega$, where Ω denotes a vector in feature space $\mathbb{F}_\kappa$, satisfying $\forall x \in \mathscr{X}$, $\Omega^T \varphi(x) = \langle f, \kappa(x,.) \rangle_{\mathscr{H}_k}$. Therefore, in feature space the model (4.9) becomes

$$\hat{y}_k = \Omega^T \varphi(x_k) \qquad (4.11)$$

This is a linear model in feature space, with $\varphi(x_k)$ as the input, and Ω as the weight vector. It is worth noting that the model (4.11) is actually a nonlinear model

---

[3] The Mercer theorem states that any reproducing kernel $\kappa(x, x')$ can be expanded as follows [160]:

$$\kappa(x, x') = \sum_{i=1}^{\infty} \lambda_i \phi_i(x) \phi_i(x')$$

where λ_i and ϕ_i are the eigenvalues and the eigenfunctions, respectively. The eigenvalues are nonnegative. Therefore, a mapping φ can be constructed as

$$\varphi : \mathscr{X} \mapsto \mathbb{F}_\kappa$$
$$\varphi(x) = \left[\sqrt{\lambda_1} \phi_1(x), \sqrt{\lambda_2} \phi_2(x), \dots \right]^T$$

By construction, the dimensionality of $\mathbb{F}_\kappa$ is determined by the number of strictly positive eigenvalues, which can be infinite (e.g., for the Gaussian kernel case).

in input space, since the mapping φ is in general a nonlinear mapping. The key principle behind kernel method is that, as long as a linear model (or algorithm) in high-dimensional feature space can be formulated in terms of inner products, a nonlinear model (or algorithm) can be obtained by simply replacing the inner product with a Mercer kernel. The model (4.11) can also be regarded as a "parameterized" model in feature space, where the parameter is the weight vector $\Omega \in \mathbb{F}_\kappa$.

4.1.2 Criterion Function

The criterion (risk or cost) function in system identification reflects how well the model fits the experimental data. In most cases, the criterion is a functional of the identification error e_k, with the form

$$R = E[l(e_k)] \tag{4.12}$$

where $l(.)$ is a loss function, which usually satisfies

Nonnegativity: $l(e) \geq 0$;
Symmetry: $l(-e) = l(e)$;
Monotonicity: $\forall |e_1| > |e_2|,\ l(e_1) \geq l(e_2)$;
Integrability: i.e., $l(.)$ is an integrable function.

Typical examples of criterion (4.12) include the mean square error (MSE), mean absolute deviation (MAD), mean p-power error (MPE), and so on. In practice, the error distribution is in general unknown, and hence, we have to estimate the expectation value in Eq. (4.12) using sample data. The estimated criterion function is called the empirical criterion function (empirical risk). Given a loss function $l(.)$, the empirical criterion function $\hat{R}$ can be computed as follows:

a. Instantaneous criterion function: $\hat{R} = l(e_k)$;
b. Average criterion function: $\hat{R} = \frac{1}{N} \sum_{k=1}^{N} l(e_k)$;
c. Weighted average criterion function: $\hat{R} = \frac{1}{N} \sum_{k=1}^{N} \gamma_k l(e_k)$.

Note that for MSE criterion ($l(e) = e^2$), the average criterion function is the well-known least-squares criterion function (sum of the squared errors).

Besides the criterion functions of form (4.12), there are many other criterion functions for system identification. In this chapter, we will discuss system identification under MEE criterion.

4.1.3 Identification Algorithm

Given a parameterized model, the identification error e_k can be expressed as a function of the parameters. For example, for the linear-in-parameter model (4.1), we have

$$e_k = z_k - h_k^T W \tag{4.13}$$

which is a linear function of W (assuming z_k and h_k are known). Similarly, the criterion function R (or the empirical criterion function $\hat{R}$) can also be expressed as a function of the parameters, denoted by $R(W)$ (or $\hat{R}(W)$). Therefore, the identification criterion represents a hyper-surface in the parameter space, which is called the *performance surface*.

The parameter W can be identified through searching the optima (minima or maxima) of the performance surface. There are two major ways to do this. One is the batch mode and the other is the online (sequential) mode.

4.1.3.1 Batch Identification

In batch mode, the identification of parameters is done only after collecting a number of samples or even possibly the whole training data. When these data are available, one can calculate the empirical criterion function $\hat{R}(W)$ based on the model structure. And then, the parameter W can be estimated by solving the following optimization problem:

$$\hat{W} = \arg \min_{W \in \Omega_W} \hat{R}(W) \tag{4.14}$$

where Ω_W denotes the set of all possible values of W. Sometimes, one can achieve an analytical solution by setting the gradient[4] of $\hat{R}(W)$ to zero, i.e.,

$$\frac{\partial}{\partial W} \hat{R}(W) = 0 \tag{4.15}$$

For example, with the linear-in-parameter model (4.1) and under the least-squares criterion (empirical MSE criterion), we have

$$\hat{R}(W) = \frac{1}{N} \sum_{k=1}^{N} e_k^2$$

$$= \frac{1}{N} \sum_{k=1}^{N} (z_k - h_k^T W)^2 \tag{4.16}$$

$$= \frac{1}{N} (z_N - H_N W)^T (z_N - H_N W)$$

where $z_N = [z_1, z_2, \ldots, z_N]^T$ and $H_N = [h_1, h_2, \ldots, h_N]^T$. And hence,

$$\frac{\partial}{\partial W} \{ (z_N - H_N W)^T (z_N - H_N W) \} = 0$$

$$\Rightarrow (H_N^T H_N) W = H_N^T z_N \tag{4.17}$$

---

[4] See Appendix H for the calculation of the gradient in vector or matrix form.

If $H_N^T H_N$ is a nonsingular matrix, we have

$$\hat{W} = (H_N^T H_N)^{-1} H_N^T z_N \tag{4.18}$$

In many situations, however, there is no analytical solution for $\hat{W}$, and we have to rely on nonlinear optimization techniques, such as gradient descent methods, simulated annealing methods, and genetic algorithms (GAs).

The batch mode approach has some shortcomings: (i) it is not suitable for online applications, since the identification is performed only after a number of data are available and (ii) the memory and computational requirements will increase dramatically with the increasing amount of data.

4.1.3.2 Online Identification

The online mode identification is also referred to as the sequential or incremental identification, or adaptive filtering. Compared with the batch mode identification, the sequential identification has some desirable features: (i) the training data (examples or observations) are sequentially (one by one) presented to the identification procedure; (ii) at any time, only few (usually one) training data are used; (iii) a training observation can be discarded as long as the identification procedure for that particular observation is completed; and (iv) it is not necessary to know how many total training observations will be presented. In this book, our focus is primarily on the sequential identification.

The sequential identification is usually performed by means of iterative schemes of the type

$$W_k = W_{k-1} + \Delta W_k \tag{4.19}$$

where W_k denotes the estimated parameter at k instant (iteration) and ΔW_k denotes the adjustment (correction) term. In the following, we present several simple online identification (adaptive filtering) algorithms.

4.1.3.3 Recursive Least Squares Algorithm

Given a linear-in-parameter model, the Recursive Least Squares (RLS) algorithm recursively finds the least-squares solution of Eq. (4.18). With a sequence of observations $\{h_i, z_i\}_{i=1}^{k-1}$ up to and including time $k - 1$, the least-squares solution is

$$W_{k-1} = (H_{k-1}^T H_{k-1})^{-1} H_{k-1}^T z_{k-1} \tag{4.20}$$

When a new observation $\{h_k, z_k\}$ becomes available, the parameter estimate W_k is

$$W_k = (H_k^T H_k)^{-1} H_k^T z_k \tag{4.21}$$

One can derive the following relation between W_k and W_{k-1}:

$$W_k = W_{k-1} + G_k e_k \tag{4.22}$$

where e_k is the prediction error,

$$e_k = z_k - \boldsymbol{h}_k^T W_{k-1} \tag{4.23}$$

and G_k is the gain vector, computed as

$$G_k = \frac{P_{k-1}\boldsymbol{h}_k}{1 + \boldsymbol{h}_k^T P_{k-1}\boldsymbol{h}_k} \tag{4.24}$$

where the matrix P can be calculated recursively as follows:

$$P_k = P_{k-1} - G_k \boldsymbol{h}_k^T P_{k-1} \tag{4.25}$$

Equations (4.22)−(4.25) constitute the RLS algorithm.

Compared to most of its competitors, the RLS exhibits very fast convergence. However, this benefit is achieved at the cost of high computational complexity. If the dimension of $\boldsymbol{h}_k$ is m, then the time and memory complexities of RLS are both $O(m^2)$.

4.1.3.4 Least Mean Square Algorithm

The Least Mean Square (LMS) algorithm is much simpler than RLS, which is a stochastic gradient descent algorithm under the instantaneous MSE cost $J(k) = \frac{e_k^2}{2}$. The weight update equation for LMS can be simply derived as follows:

$$\begin{aligned} W_k &= W_{k-1} - \eta \left[\frac{\partial}{\partial W} \frac{e_k^2}{2} \right]_{W_{k-1}} \\ &= W_{k-1} - \eta e_k \left[\frac{\partial}{\partial W} (z_k - \hat{y}_k) \right]_{W_{k-1}} \\ &= W_{k-1} + \eta e_k \left[\frac{\partial \hat{y}_k}{\partial W} \right]_{W_{k-1}} \end{aligned} \tag{4.26}$$

where $\eta > 0$ is the step-size (adaptation gain, learning rate, etc.),[5] and the term $\partial \hat{y}_k / \partial W$ is the instantaneous gradient of the model output with respect to the weight vector, whose form depends on the model structure. For a FIR filter (or

---

[5] The step-size is critical to the performance of the LMS. In general, the choice of step-size is a trade-off between the convergence rate and the asymptotic EMSE [19,20].

ADALINE), the instantaneous gradient will simply be the input vector X_k. In this case, the LMS algorithm becomes[6]

$$W_k = W_{k-1} + \eta e_k X_k \tag{4.27}$$

The computational complexity of the LMS (4.27) is just $O(m)$, where m is the input dimension.

If the model is an MLP network, the term $\partial \hat{y}_k / \partial W$ can be computed by back propagation (BP), which is a common method of training artificial neural networks so as to minimize the objective function [53].

There are many other stochastic gradient descent algorithms that are similar to the LMS. Typical examples include the least absolute deviation (LAD) algorithm [31] and the least mean fourth (LMF) algorithm [26]. The LMS, LAD, and LMF algorithms are all special cases of the least mean p-power (LMP) algorithm [30]. The LMP algorithm aims to minimize the p-power of the error, which can be derived as

$$
\begin{aligned}
W_k &= W_{k-1} - \eta \left[\tfrac{\partial}{\partial W} |e_k|^p \right]_{W_{k-1}} \\
&= W_{k-1} - p\eta |e_k|^{p-1} \operatorname{sign}(e_k) \left[\tfrac{\partial}{\partial W} (z_k - \hat{y}_k) \right]_{W_{k-1}} \\
&= W_{k-1} + p\eta |e_k|^{p-1} \operatorname{sign}(e_k) \left[\tfrac{\partial \hat{y}_k}{\partial W} \right]_{W_{k-1}}
\end{aligned}
\tag{4.28}
$$

For the cases $p = 1, 2, 4$, the above algorithm corresponds to the LAD, LMS, and LMF algorithms, respectively.

4.1.3.5 Kernel Adaptive Filtering Algorithms

The kernel adaptive filtering (KAF) algorithms are a family of nonlinear adaptive filtering algorithms developed in kernel (or feature) space [12], by using the linear structure and inner product of this space to implement the well-established linear adaptive filtering algorithms (e.g., LMS, RLS, etc.) and to obtain nonlinear filters in the original input space. They have several desirable features: (i) if choosing a universal kernel (e.g., Gaussian kernel), they are universal approximators; (ii) under MSE criterion, the performance surface is quadratic in feature space so gradient descent learning does not suffer from local minima; and (iii) if pruning the redundant features, they have moderate complexity in terms of computation and memory. Typical KAF algorithms include the kernel recursive least squares (KRLS) [176], kernel least mean square (KLMS) [177], kernel affine projection algorithms (KAPA) [178], and so on. When the kernel is radial (such as the Gaussian kernel), they naturally build a growing RBF network, where the weights are directly related

---

[6] The LMS algorithm usually assumes an FIR model [19,20].

to the errors at each sample. In the following, we only discuss the KLMS algorithm. Interesting readers can refer to Ref. [12] for further information about KAF algorithms.

Let X_k be an m-dimensional input vector. We can transform X_k into a high-dimensional feature space $\mathbb{F}_\kappa$ (induced by kernel κ) through a nonlinear mapping φ, i.e., $\varphi_k = \varphi(X_k)$. Suppose the model in feature space is given by Eq. (4.11). Then using the LMS algorithm on the transformed observation sequence $\{\varphi_k, z_i\}$ yields [177]

$$\begin{cases} \Omega_0 = 0 \\ e_k = z_k - \Omega_{k-1}{}^T \varphi_k \\ \Omega_k = \Omega_{k-1} + \eta e_k \varphi_k \end{cases} \tag{4.29}$$

where Ω_k denotes the estimated weight vector (at iteration k) in feature space. The KLMS (4.29) is very similar to the LMS algorithm, except for the dimensionality (or richness) of the projection space. The learned mapping (model) at iteration k is the composition of Ω_k and φ, i.e., $f_k = \Omega_k{}^T \varphi(.)$. If identifying $\varphi_k = \kappa(X_k, .)$, we obtain the sequential learning rule in the original input space:

$$\begin{cases} f_0 = 0 \\ e_k = z_k - f_{k-1}(X_k) \\ f_k = f_{k-1} + \eta e_k \kappa(X_k, .) \end{cases} \tag{4.30}$$

At iteration k, given an input X, the output of the filter is

$$f_k(X) = \eta \sum_{j=1}^{k} e_j \kappa(X_j, X) \tag{4.31}$$

From Eq. (4.31) we see that, if choosing a radial kernel, the KLMS produces a growing RBF network by allocating a new kernel unit for every new example with input X_k as the center and ηe_k as the coefficient. The algorithm of KLMS is summarized in Table 4.1, and the corresponding network topology is illustrated in Figure 4.2.

Selecting a proper Mercer kernel is crucial for all kernel methods. In KLMS, the kernel is usually chosen to be a normalized Gaussian kernel:

$$\kappa(x, x') = \exp\left(-\frac{1}{2\sigma^2}||x - x'||^2\right) = \exp(-\zeta||x - x'||^2) \tag{4.32}$$

where $\sigma > 0$ is the kernel size (kernel width) and $\zeta = 1/2\sigma^2$ is called the kernel parameter. The kernel size in Gaussian kernel is an important parameter that controls the degree of smoothing and consequently has significant influence on the

Table 4.1 The KLMS Algorithm

Initialization:

Choosing Mercer kernel κ and step-size η
$\alpha_1 = \eta z_1,\quad \boldsymbol{C}(1) = \{X_1\},\quad f_1 = \alpha_1 \kappa(X_1, .)$
Computation:
while$\{X_k, z_k\}$ ($k > 1$) available **do**
1. Compute the filter output: $f_{k-1}(X_k) = \sum_{j=1}^{k-1} \alpha_j \kappa(X_j, X_k)$
2. Compute the prediction error: $e_k = z_k - f_{k-1}(X_k)$
3. Store the new center: $\boldsymbol{C}(k) = \{\boldsymbol{C}(k-1), X_k\}$
4. Compute and store the coefficients: $\alpha_k = \eta e_k$
end while

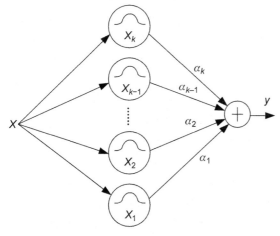

Figure 4.2 Network topology of KLMS at iteration k.

learning performance. Usually, the kernel size can be set manually or estimated in advance by Silverman's rule [97]. The role of the step-size in KLMS remains in principle the same as the step-size in traditional LMS algorithm. Specifically, it controls the compromise between convergence speed and misadjustment. It has also been shown in Ref. [177] that the step-size in KLMS plays a similar role as the regularization parameter.

The main bottleneck of KLMS (as well as other KAF algorithms) is the linear growing network with each new sample, which poses both computational and memory issues especially for continuous adaptation scenarios. In order to curb the network growth and to obtain a compact representation, a variety of sparsification techniques can be applied, where only the important input data are accepted as the centers. Typical sparsification criteria include the novelty criterion [179], approximate linear dependency (ALD) criterion [176], coherence criterion [180], surprise criterion [181], and so on. The idea of quantization can also be used to yield a compact network with desirable accuracy [182].

4.2   MEE Identification Criterion

Most of the existing approaches to parameter identification utilized the MSE (or equivalently, Gaussian likelihood) as the identification criterion function. The MSE is mathematically tractable and under Gaussian assumption is an optimal criterion for linear system. However, it is well known that MSE may be a poor descriptor of optimality for nonlinear and non-Gaussian (e.g., multimodal, heavy-tail, or finite range distributions) situations, since it constrains only the second-order statistics. To address this issue, one can select some criterion beyond second-order statistics that does not suffer from the limitation of Gaussian assumption and can improve performance in many realistic scenarios. Information theoretic quantities (entropy, divergence, mutual information, etc.) as identification criteria attract ever-increasing attention to this end, since they can capture higher order statistics and information content of signals rather than simply their energy [64]. In the following, we discuss the MEE criterion for system identification.

Under MEE criterion, the parameter vector (weight vector) W can be identified by solving the following optimization problem:

$$
\begin{aligned}
\hat{W} &= \arg\min_{W \in \Omega_W} H(e_k) \\
&= \arg\min_{W \in \Omega_W} -\int_{-\infty}^{\infty} p_e(\xi) \log p_e(\xi) d\xi \\
&= \arg\min_{W \in \Omega_W} E_e[-\log p_e(e_k)]
\end{aligned}
\tag{4.33}
$$

where $p_e(.)$ denotes the probability density function (PDF) of error $e_k = z_k - \hat{y}_k$. If using the order-α Renyi entropy ($\alpha > 0$, $\alpha \neq 1$) of the error as the criterion function, the estimated parameter will be

$$
\begin{aligned}
\hat{W} &= \arg\min_{W \in \Omega_W} H_\alpha(e_k) \\
&= \arg\min_{W \in \Omega_W} \frac{1}{1-\alpha} \log \int_{-\infty}^{\infty} p_e^\alpha(\xi) d\xi \\
&= \arg\min_{W \in \Omega_W} \frac{1}{1-\alpha} \log V_\alpha(e_k) \\
&\stackrel{(a)}{=}
\begin{cases}
\arg\min_{W \in \Omega_W} V_\alpha(e_k) & \text{if} \quad \alpha < 1 \\
\arg\max_{W \in \Omega_W} V_\alpha(e_k) & \text{if} \quad \alpha > 1
\end{cases}
\end{aligned}
\tag{4.34}
$$

where (a) follows from the monotonicity of logarithm function, $V_\alpha(e_k)$ is the order-α information potential (IP) of the error e_k. If $\alpha < 1$, minimizing the order-α Renyi entropy is equivalent to minimizing the order-α IP; while if $\alpha > 1$, minimizing the order-α Renyi entropy is equivalent to maximizing the order-α IP. In practical application, we often use the order-α IP instead of the order-α Renyi entropy as the criterion function for identification.

Further, if using the ϕ-entropy of the error as the criterion function, we have

$$
\begin{aligned}
\hat{W} &= \arg\min_{W \in \Omega_W} H_\phi(e_k) \\
&= \arg\min_{W \in \Omega_W} \left\{ \int_{-\infty}^{\infty} \phi[p_e(\xi)]d\xi \right\} \\
&= \arg\min_{W \in \Omega_W} E_e[\psi(p_e(e_k))]
\end{aligned}
\tag{4.35}
$$

where $\psi(x) = \phi(x)/x$. Note that the ϕ-entropy criterion includes the Shannon entropy ($\phi(x) = -x \log x$) and order-α information potential ($\phi(x) = \text{sign}(1 - \alpha)x^\alpha$) as special cases.

The error entropy is a functional of error distribution. In practice, the error distribution is usually unknown to us, and so is the error entropy. And hence, we have to estimate the error entropy from error samples, and use the estimated error entropy (called the empirical error entropy) as a criterion to identify the system parameter. In the following, we present several common approaches to estimating the entropy from sample data.

4.2.1 Common Approaches to Entropy Estimation

A straight way to estimate the entropy is to estimate the underlying distribution based on available samples, and plug the estimated distributions into the entropy expression to obtain the entropy estimate (the so-called "*plug-in approach*") [183]. Several plug-in estimates of the Shannon entropy (extension to other entropy definitions is straightforward) are presented as follows.

4.2.1.1 Integral Estimate

Denote $\hat{p}_N(x)$ the estimated PDF based on sample $S_N = \{x_1, x_2, \ldots, x_N\}$. Then the integral estimate of entropy is of the form

$$
H_N = -\int_{A_N} \hat{p}_N(x) \log \hat{p}_N(x) dx
\tag{4.36}
$$

where A_N is a set typically used to exclude the small or tail values of $\hat{p}_N(x)$. The evaluation of Eq. (4.36) requires numerical integration and is not an easy task in general.

4.2.1.2 Resubstitution Estimate

The resubstitution estimate substitutes the estimated PDF into the sample mean approximation of the entropy measure (approximating the expectation value by its sample mean), which is of the form

$$H_N = -\frac{1}{N}\sum_{i=1}^{N} \log \hat{p}_N(x_i) \tag{4.37}$$

This estimation method is considerably simpler than the integral estimate, since it involves no numerical integration.

4.2.1.3 Splitting Data Estimate

Here, we decompose the sample $S_N = \{x_1, x_2, \ldots, x_N\}$ into two sub samples: $S_L = \{x_1, \ldots, x_L\}$, $S_M = \{x_1^*, \ldots, x_M^*\}$, $N = L + M$. Based on subsample S_L, we obtain a density estimate $\hat{p}_L(x)$, and then, using this density estimate and the second subsample S_M, we estimate the entropy by

$$H_N = -\frac{1}{M}\sum_{i=1}^{M} \mathbb{I}[x_i^* \in A_L]\log \hat{p}_L(x_i^*) \tag{4.38}$$

where $\mathbb{I}[.]$ is the indicator function and the set $A_L = \{x : \hat{p}_L(x) \geq a_L\}$ $(0 < a_L \rightarrow 0)$. The splitting data estimate is different from the resubstitution estimate in that it uses different samples to estimate the density and to calculate the sample mean.

4.2.1.4 Cross-validation Estimate

If $\hat{p}_{N,i}$ denotes a density estimate based on sample $S_{N,i} = S_N - \{x_i\}$ (i.e., leaving x_i out), then the cross-validation estimate of entropy is

$$H_N = -\frac{1}{N}\sum_{i=1}^{N} I[x_i \in A_{N,i}]\log \hat{p}_{N,i}(x_i) \tag{4.39}$$

A key step in plug-in estimation is to estimate the PDF from sample data. In the literature, there are mainly two approaches for estimating the PDF of a random variable based on its sample data: parametric and nonparametric. Accordingly, there are also parametric and nonparametric entropy estimations. The parametric density estimation assumes a parametric model of the density and estimates the involved parameters using classical estimation methods like the maximum likelihood (ML) estimation. This approach needs to select a suitable parametric model of the density, which depends upon some prior knowledge. The nonparametric density estimation, however, does not need to select a parametric model, and can estimate the PDF of any distribution.

The histogram density estimation (HDE) and kernel density estimation (KDE) are two popular nonparametric density estimation methods. Here we only discuss the KDE method (also referred to as Parzen window method), which has been widely used in nonparametric regression and pattern recognition. Given a set of independent and identically distributed (i.i.d.) samples[7] $\{x_1, \ldots, x_N\}$ drawn from $p(x)$, the KDE for $p(x)$ is given by [97]

$$\hat{p}_N(x) = \frac{1}{N} \sum_{i=1}^{N} K_h(x - x_i) \tag{4.40}$$

where $K_h(.)$ denotes a kernel function[8] with width h, satisfying the following conditions:

$$\begin{cases} K_h(x) \geq 0 \\ \int_{-\infty}^{+\infty} K_h(x) = 1 \\ K_h(x) = K(x/h)/h \end{cases} \tag{4.41}$$

where $K(.)$ is the kernel function with width 1. To make the estimated PDF smooth, the kernel function is usually selected to be a continuous and differentiable (and preferably symmetric and unimodal) function. The most widely used kernel function in KDE is the Gaussian function:

$$K_h(x) = \frac{1}{\sqrt{2\pi}h} \exp\left(-\frac{x^2}{2h^2}\right) \tag{4.42}$$

The kernel width of the Gaussian kernel can be optimized by the ML principle, or selected according to rules-of-thumb, such as Silverman's rule [97].

With a fixed kernel width h, we have

$$\lim_{N \to \infty} \hat{p}_N(x) = p(x) * K_h(x) \tag{4.43}$$

where $*$ denotes the convolution operator. Using a suitable annealing rate for the kernel width, the KDE can be asymptotically unbiased and consistent. Specifically, if $\lim_{N \to \infty} h_N = 0$ and $\lim_{N \to \infty} Nh_N = \infty$, then $\lim_{N \to \infty} \hat{p}_N(x) = p(x)$ in probability [98].

In addition to the plug-in methods described previously, there are many other methods for entropy estimation, such as the sample-spacing method and the nearest neighbor method. In the following, we derive the sample-spacing estimate. First, let us express the Shannon entropy as [184]

$$H(p) = \int_0^1 \log\left(\frac{\partial}{\partial P} F^{-1}(P)\right) dP \tag{4.44}$$

---

[7] In practical applications, if the samples are not i.i.d., the KDE method can still be applied.
[8] The kernel function for density estimation is not necessarily a Mercer kernel. In this book, to make a distinction between these two types of kernels, we denote them by K and κ, respectively.

where $\frac{\partial}{\partial x}F(x) = p(x)$. Using the slope of the curve $F^{-1}(P)$ to approximate the derivative, we have

$$\frac{\partial}{\partial P}F^{-1}(P) \approx \frac{(x_{N,i+m} - x_{N,i})}{m/N} \tag{4.45}$$

where $x_{N,1} < x_{N,2} < \cdots < x_{N,N}$ is the order statistics of the sample $S_N = \{x_1, x_2, \ldots, x_N\}$, and $(x_{N,i+m} - x_{N,i})$ is the m-order sample spacing $(1 \leq i < i + m \leq N)$. Hence, the sample-spacing estimate of the entropy is

$$H_{m,N} = \frac{1}{N}\sum_{i=1}^{N-m} \log\left(\frac{N}{m}(x_{N,i+m} - x_{N,i})\right) \tag{4.46}$$

If adding a correction term to compensate the asymptotic bias, we get

$$H_{m,N} = \frac{1}{N}\sum_{i=1}^{N-m} \log\left(\frac{N}{m}(x_{N,i+m} - x_{N,i})\right) + \psi(m) + \log m \tag{4.47}$$

where $\psi(x) = (\log\Gamma(x))'$ is the Digamma function ($\Gamma(.)$ is the Gamma function).

4.2.2 Empirical Error Entropies Based on KDE

To calculate the empirical error entropy, we usually adopt the resubstitution estimation method with error PDF estimated by KDE. This approach has some attractive features: (i) it is a nonparametric method, and hence requires no prior knowledge on the error distribution; (ii) it is computationally simple, since no numerical integration is needed; and (iii) if choosing a smooth and differentiable kernel function, the empirical error entropy (as a function of the error sample) is also smooth and differentiable (this is very important for the calculation of the gradient).

Suppose now a set of error samples $S_e = \{e_1, e_2, \ldots, e_N\}$ are available. By KDE, the error density can be estimated as

$$\hat{p}_N(e) = \frac{1}{N}\sum_{i=1}^{N} K_h(e - e_i) \tag{4.48}$$

Then by resubstitution estimation method, we obtain the following empirical error entropies:

1. Empirical Shannon entropy

$$\hat{H}(e) = -\frac{1}{N}\sum_{j=1}^{N} \log\left(\frac{1}{N}\sum_{i=1}^{N} K_h(e_j - e_i)\right) \tag{4.49}$$

2. Empirical order-α Renyi entropy

$$\hat{H}_\alpha(e) = \frac{1}{1-\alpha}\log \hat{V}_\alpha(e) = \frac{1}{1-\alpha}\log\left\{\frac{1}{N}\sum_{j=1}^{N}\left(\frac{1}{N}\sum_{i=1}^{N}K_h(e_j-e_i)\right)^{\alpha-1}\right\} \qquad (4.50)$$

where $\hat{V}_\alpha(e)$ is the empirical order-α IP, i.e.,

$$\hat{V}_\alpha(e) = \frac{1}{N}\sum_{j=1}^{N}\left(\frac{1}{N}\sum_{i=1}^{N}K_h(e_j-e_i)\right)^{\alpha-1} \qquad (4.51)$$

3. Empirical ϕ-entropy

$$\hat{H}_\phi(e) = \frac{1}{N}\sum_{j=1}^{N}\psi\left(\frac{1}{N}\sum_{i=1}^{N}K_h(e_j-e_i)\right) \qquad (4.52)$$

It is worth noting that for quadratic information potential (QIP) ($\alpha = 2$), if choosing Gaussian kernel function, the resubstitution estimate will be identical to the integral estimate but with kernel width $\sqrt{2}h$ instead of h. Specifically, if K_h is given by Eq. (4.42), we can derive

$$
\begin{aligned}
&\int_{-\infty}^{\infty}\left(\frac{1}{N}\sum_{i=1}^{N}K_h(e-e_i)\right)^2 de \\
&= \frac{1}{N^2}\int_{-\infty}^{\infty}\left(\sum_{j=1}^{N}\sum_{i=1}^{N}K_h(e-e_j)K_h(e-e_i)\right)de \\
&= \frac{1}{N^2}\sum_{j=1}^{N}\sum_{i=1}^{N}\int_{-\infty}^{\infty}K_h(e-e_j)K_h(e-e_i)de \\
&= \frac{1}{N^2}\sum_{j=1}^{N}\sum_{i=1}^{N}K_{\sqrt{2}h}(e_j-e_i)
\end{aligned}
\qquad (4.53)
$$

This result comes from the fact that the integral of the product of two Gaussian functions can be exactly evaluated as the value of the Gaussian function computed at the difference of the arguments and whose variance is the sum of the variances of the two original Gaussian functions. From Eq. (4.53), we also see that the QIP can be simply calculated by the double summation over error samples. Due to this fact, when using order-α IP, we usually set $\alpha = 2$.

In the following, we present some important properties of the empirical error entropy, and our focus is mainly on the order-α Renyi entropy (or order-α IP) [64,67].

Property 1: $\hat{H}_\alpha(e+c) = \hat{H}_\alpha(e)$.

Proof: Let $\tilde{e} = e + c$, where $c \in \mathbb{R}$ is an arbitrary constant, $\tilde{e}_i = e_i + c, i = 1, 2, \ldots, N$, then we have

$$
\begin{aligned}
\hat{H}_\alpha(\tilde{e}) &= \frac{1}{1-\alpha}\log\left\{\frac{1}{N^\alpha}\sum_{j=1}^{N}\left(\sum_{i=1}^{N}K_h(\tilde{e}_j - \tilde{e}_i)\right)^{\alpha-1}\right\} \\
&= \frac{1}{1-\alpha}\log\left\{\frac{1}{N^\alpha}\sum_{j=1}^{N}\left(\sum_{i=1}^{N}K_h((e_j+c)-(e_i+c))\right)^{\alpha-1}\right\} \\
&= \frac{1}{1-\alpha}\log\left\{\frac{1}{N^\alpha}\sum_{j=1}^{N}\left(\sum_{i=1}^{N}K_h(e_j - e_i)\right)^{\alpha-1}\right\} \\
&= \hat{H}_\alpha(e)
\end{aligned}
\tag{4.54}
$$

Remark: Property 1 shows that the empirical error entropy is also shift-invariant. In system identification with MEE criterion, we usually set a bias at the model output to achieve a zero-mean error.

Property 2: $\lim\limits_{\alpha \to 1}\hat{H}_\alpha(e) = \hat{H}(e)$, where $\hat{H}(e) = -\frac{1}{N}\sum\limits_{j=1}^{N}\log\left(\frac{1}{N}\sum\limits_{i=1}^{N}K_h(e_j - e_i)\right)$ is the empirical Shannon entropy.

Proof:

$$
\begin{aligned}
\lim_{\alpha \to 1}\hat{H}_\alpha(e) &= \lim_{\alpha \to 1}\frac{1}{1-\alpha}\log\left(\frac{1}{N}\sum_{j=1}^{N}\left(\frac{1}{N}\sum_{i=1}^{N}K_h(e_j - e_i)\right)^{\alpha-1}\right) \\
&= \frac{\lim\limits_{\alpha \to 1}\dfrac{\partial}{\partial\alpha}\left\{\log\left(\dfrac{1}{N}\sum\limits_{j=1}^{N}\left(\dfrac{1}{N}\sum\limits_{i=1}^{N}K_h(e_j - e_i)\right)^{\alpha-1}\right)\right\}}{\lim\limits_{\alpha \to 1} -1} \\
&= -\lim_{\alpha \to 1}\frac{\left(\dfrac{1}{N}\sum\limits_{j=1}^{N}\left(\dfrac{1}{N}\sum\limits_{i=1}^{N}K_h(e_j - e_i)\right)^{\alpha-1}\log\left(\dfrac{1}{N}\sum\limits_{i=1}^{N}K_h(e_j - e_i)\right)\right)}{\left(\dfrac{1}{N}\sum\limits_{j=1}^{N}\left(\dfrac{1}{N}\sum\limits_{i=1}^{N}K_h(e_j - e_i)\right)^{\alpha-1}\right)} \\
&= -\frac{1}{N}\sum_{j=1}^{N}\log\left(\frac{1}{N}\sum_{i=1}^{N}K_h(e_j - e_i)\right) = \hat{H}(e)
\end{aligned}
$$

$$\tag{4.55}$$

Property 3: Let's denote $\hat{H}_\alpha(e) = \hat{H}_{\alpha,h}(e)$ (h is the kernel width). Then $\forall c \in \mathbb{R}$, $c \neq 0$, we have $\hat{H}_{\alpha,|c|h}(ce) = \hat{H}_{\alpha,h}(e) + \log|c|$.

Proof:

$$
\begin{aligned}
\hat{H}_{\alpha,|c|h}(ce) &= \frac{1}{1-\alpha} \log \left(\frac{1}{N} \sum_{j=1}^{N} \left(\frac{1}{N} \sum_{i=1}^{N} K_{|c|h}(ce_j - ce_i) \right)^{\alpha-1} \right) \\
&\overset{(a)}{=} \frac{1}{1-\alpha} \log \left(\frac{1}{N} \sum_{j=1}^{N} \left(\frac{1}{N} \sum_{i=1}^{N} \frac{1}{|c|} K_h \left(\frac{ce_j - ce_i}{c} \right) \right)^{\alpha-1} \right) \\
&= \frac{1}{1-\alpha} \log \left(\frac{1}{N|c|^{\alpha-1}} \sum_{j=1}^{N} \left(\frac{1}{N} \sum_{i=1}^{N} K_h(e_j - e_i) \right)^{\alpha-1} \right) \\
&= \hat{H}_{\alpha,h}(e) + \log|c|
\end{aligned}
\tag{4.56}
$$

where (a) is because that $\forall c > 0$, the kernel function $K_h(x)$ satisfies $K_{ch}(x) = K_h(x/c)/c$.

Property 4: $\lim_{N \to \infty} \hat{H}_\alpha(e) \to H_\alpha(\hat{e}) \geq H_\alpha(e)$, where $\hat{e}$ is a random variable with PDF $p_e * K_h$ (* denotes the convolution operator).

Proof: According to the theory of KDE [97,98], we have

$$
\lim_{N \to \infty} \hat{p}_N \to p_e * K_h
\tag{4.57}
$$

Hence, $\lim_{N \to \infty} \hat{H}_\alpha(e) \to H_\alpha(\hat{e})$. Since the PDF of the sum of two independent random variables is equal to the convolution of their individual PDFs, $\hat{e}$ can be considered as the sum of the error e and another random variable that is independent of the error and has PDF K_h. And since the entropy of the sum of two independent random variables is no less than the entropy of each individual variable, we have $H_\alpha(\hat{e}) \geq H_\alpha(e)$.

Remark: Property 4 implies that minimizing the empirical error entropy will minimize an upper bound of the actual error entropy. In general, an identification algorithm seeks extrema (either minimum or maximum) of the cost function, independently to its actual value, so the dependence on estimation error is decreased.

Property 5: If $K_h(0) = \max_x K_h(x)$, then $\hat{H}_\alpha(e) \geq -\log K_h(0)$, with equality if $e_1 = e_2 = \cdots = e_N$.

Proof: Consider the case where $\alpha > 1$ ($\alpha < 1$ is similar), we have

$$\hat{H}_\alpha(e) = \frac{1}{1-\alpha} \log \left\{ \frac{1}{N^\alpha} \sum_{j=1}^{N} \left(\sum_{i=1}^{N} K_h(e_j - e_i) \right)^{\alpha-1} \right\}$$

$$\geq \frac{1}{1-\alpha} \log \left\{ \frac{1}{N^\alpha} \sum_{j=1}^{N} \left(\sum_{i=1}^{N} K_h(0) \right)^{\alpha-1} \right\} \tag{4.58}$$

$$= -\log K_h(0)$$

If $\forall i,j$, $e_j - e_i = 0$, i.e., $e_1 = e_2 = \cdots = e_N$, the equality will hold.

Remark: Property 5 suggests that if the kernel function reaches its maximum value at zero, then when all the error samples have the same value, the empirical error entropy reaches minimum, and in this case the uncertainty of the error is minimum.

Property 6: If the kernel function $K_h(.)$ is continuously differentiable, symmetric, and unimodal, then the empirical error entropy is smooth at the global minimum of Property 5, that is, $\hat{H}_\alpha$ has zero-gradient and positive semidefinite Hessian matrix at $e = [e_1, \ldots, e_N]^T = \mathbf{0}$.

Proof:

$$\begin{cases} \dfrac{\partial \hat{H}_\alpha}{\partial e_k} = \dfrac{1}{1-\alpha} \dfrac{\partial \hat{V}_\alpha / \partial e_k}{\hat{V}_\alpha} \\[4mm] \dfrac{\partial^2 \hat{H}_\alpha}{\partial e_l \partial e_k} = \dfrac{1}{1-\alpha} \dfrac{(\partial^2 \hat{V}_\alpha / \partial e_l \partial e_k) - (\partial \hat{V}_\alpha / \partial e_k)(\partial \hat{V}_\alpha / \partial e_l)}{\hat{V}_\alpha^2} \end{cases} \tag{4.59}$$

If $e = \mathbf{0}$, we can calculate

$$\begin{cases} \hat{V}_\alpha|_{e=0} = K_h^{\alpha-1}(0) \\[3mm] \dfrac{\partial \hat{V}_\alpha}{\partial e_k}\Big|_{e=0} = \dfrac{(\alpha-1)}{N^\alpha} \left[N^{\alpha-1} K_h^{\alpha-2}(0) K'_h(0) - N^{\alpha-1} K_h^{\alpha-2}(0) K'_h(0) \right] = 0 \\[3mm] \dfrac{\partial^2 \hat{V}_\alpha}{\partial^2 e_k}\Big|_{e=0} = \dfrac{(\alpha-1)(N-1)K_h^{\alpha-3}(0)}{N^2} \left[(\alpha-2)K_h'^2(0) + 2K_h(0)K_h''(0) \right] \\[3mm] \dfrac{\partial^2 \hat{V}_\alpha}{\partial e_l \partial e_k}\Big|_{e=0} = -\dfrac{(\alpha-1)K_h^{\alpha-3}(0)}{N^2} \left[(\alpha-2)K_h'^2(0) + 2K_h(0)K_h''(0) \right] \end{cases} \tag{4.60}$$

Hence, the gradient vector $(\partial \hat{H}_\alpha / \partial e)|_{e=0} = 0$, and the Hessian matrix is

$$\frac{\partial^2 \hat{H}_\alpha}{\partial e_l \partial e_k}\Big|_{e=0} = \begin{cases} -(N-1)K_h^{-\alpha-1}(0)[(\alpha-2)K_h'^2(0) + 2K_h(0)K_h''(0)]/N^2, & l = k \\ K_h^{-\alpha-1}(0)[(\alpha-2)K_h'^2(0) + 2K_h(0)K_h''(0)]/N^2, & l \neq k \end{cases}$$

(4.61)

whose eigenvalue−eigenvector pairs are

$$\{0, [1, \ldots, 1]^T\}, \quad \{aN, [1, -1, 0, \ldots, 0]^T\}, \quad \{aN, [1, 0, -1, \ldots, 0]^T\}, \ldots \quad (4.62)$$

where $a = -K_h^{-\alpha-1}(0)[(\alpha-2)K_h'^2(0) + 2K_h(0)K_h''(0)]/N^2$. According to the assumptions we have $a \geq 0$, therefore this Hessian matrix is positive semidefinite.

Property 7: With Gaussian kernel, the empirical QIP $\hat{V}_2(e)$ can be expressed as the squared norm of the mean vector of the data in kernel space.

Proof: The Gaussian kernel is a Mercer kernel, and can be written as an inner product in the kernel space (RKHS):

$$K_h(e_i - e_j) = \langle \varphi(e_i), \varphi(e_j) \rangle_{\mathscr{H}_{K_h}} \quad (4.63)$$

where $\varphi(.)$ defines the nonlinear mapping between input space and kernel space $\mathscr{H}_{K_h}$. Hence the empirical QIP can also be expressed in terms of an inner product in kernel space:

$$\begin{aligned} \hat{V}_2(e) &= \frac{1}{N^2} \sum_{i=1}^{N} \sum_{j=1}^{N} K_h(e_j - e_i) \\ &= \frac{1}{N^2} \sum_{i=1}^{N} \sum_{j=1}^{N} \langle \varphi(e_i), \varphi(e_j) \rangle_{H_{K_h}} \\ &= \left\langle \frac{1}{N} \sum_{i=1}^{N} \varphi(e_i), \frac{1}{N} \sum_{i=1}^{N} \varphi(e_i) \right\rangle_{H_{K_h}} \\ &= \mathbf{m}^T \mathbf{m} = \|\mathbf{m}\|^2 \end{aligned}$$

(4.64)

where $\mathbf{m} = \frac{1}{N} \sum_{i=1}^{N} \varphi(e_i)$ is the mean vector of the data in kernel space.

In addition to the previous properties, the literature [102] points out that the empirical error entropy has the *dilation* feature, that is, as the kernel width h increases, the performance surface (surface of the empirical error entropy in parameter space) will

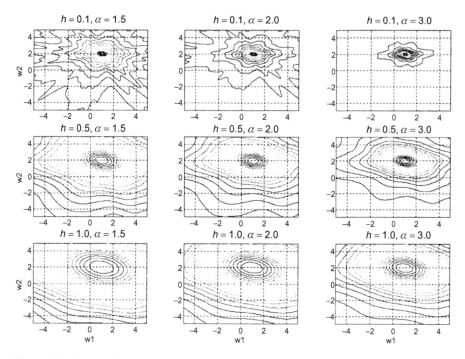

Figure 4.3 Contour plots of a two-dimensional performance surface of ADALINE parameter identification based on the order-α IP criterion (adopted from Ref. [102]).

become more and more flat, thus leading to the local extrema reducing gradually and even disappearing. Figure 4.3 illustrates the contour plots of a two-dimensional performance surface of ADALINE parameter identification based on the order-α IP criterion. It is clear to see that the IPs corresponding to different α values all have the feature of dilation. The dilation feature implies that one can obtain desired performance surface by means of proper selection of the kernel width.

4.3 Identification Algorithms Under MEE Criterion

4.3.1 Nonparametric Information Gradient Algorithms

In general, information gradient (IG) algorithms refer to the gradient-based identification algorithms under MEE criterion (i.e., minimizing the empirical error entropy), including the batch information gradient (BIG) algorithm, sliding information gradient algorithm, forgetting recursive information gradient (FRIG) algorithm, and stochastic information gradient (SIG) algorithm. If the empirical error entropy is estimated by nonparametric approaches (like KDE), then they are called the nonparametric IG algorithms. In the following, we present several nonparametric IG algorithms that are based on ϕ-entropy and KDE.

4.3.1.1 BIG Algorithm

With the empirical ϕ-entropy as the criterion function, the BIG identification algorithm is derived as follows:

$$W_k = W_{k-1} - \eta \frac{\partial}{\partial W} \hat{H}_\phi(e)$$

$$= W_{k-1} - \frac{\eta}{N} \frac{\partial}{\partial W} \left\{ \sum_{j=1}^{N} \psi \left(\frac{1}{N} \sum_{i=1}^{N} K_h(e_j - e_i) \right) \right\}$$

$$= W_{k-1} - \frac{\eta}{N^2} \sum_{j=1}^{N} \left\{ \psi' \left(\frac{1}{N} \sum_{i=1}^{N} K_h(e_j - e_i) \right) \sum_{i=1}^{N} \left(K'_h(e_j - e_i) \left(\frac{\partial e_j}{\partial W} - \frac{\partial e_i}{\partial W} \right) \right) \right\}$$

$$= W_{k-1} + \frac{\eta}{N^2} \sum_{j=1}^{N} \left\{ \psi' \left(\frac{1}{N} \sum_{i=1}^{N} K_h(e_j - e_i) \right) \sum_{i=1}^{N} \left(K'_h(e_j - e_i) \left(\frac{\partial \hat{y}_j}{\partial W} - \frac{\partial \hat{y}_i}{\partial W} \right) \right) \right\}$$

$$(4.65)$$

where η is the step-size, N is the number of training data, $\psi'(.)$ and $K'_h(.)$ denote, respectively, the first-order derivatives of functions $\psi(.)$ and $K_h(.)$. The gradient $\partial \hat{y}_i / \partial W$ of the model output with respect to the parameter W depends on the specific model structure. For example, if the model is an ADALINE or FIR filter, we have $\partial \hat{y}_i / \partial W = X_i$. The reason for the algorithm named as the BIG algorithm is because that the empirical error entropy is calculated based on all the training data.

Given a specific ϕ function, we obtain a specific algorithm:

1. $\phi(x) = -x \log x$ (Shannon entropy)

$$W_k = W_{k-1} - \frac{\eta}{N} \sum_{j=1}^{N} \left\{ \left(\sum_{i=1}^{N} K_h(e_j - e_i) \right)^{-1} \sum_{i=1}^{N} \left(K'_h(e_j - e_i) \left(\frac{\partial \hat{y}_j}{\partial W} - \frac{\partial \hat{y}_i}{\partial W} \right) \right) \right\} \quad (4.66)$$

2. $\phi(x) = \text{sign}(1 - \alpha)x^\alpha$ (Corresponding to order-α information potential)

$$W_k = W_{k-1} + \text{sign}(1 - \alpha) \frac{\eta(\alpha - 1)}{N^\alpha}$$

$$\times \sum_{j=1}^{N} \left\{ \left(\sum_{i=1}^{N} K_h(e_j - e_i) \right)^{\alpha-2} \sum_{i=1}^{N} \left(K'_h(e_j - e_i) \left(\frac{\partial \hat{y}_j}{\partial W} - \frac{\partial \hat{y}_i}{\partial W} \right) \right) \right\} \quad (4.67)$$

In above algorithms, if the kernel function is Gaussian function, the derivative $K'_h(e_j - e_i)$ will be

$$K'_h(e_j - e_i) = -\frac{1}{h^2} (e_j - e_i) K_h(e_j - e_i) \quad (4.68)$$

The step-size η is a crucial parameter that controls the compromise between convergence speed and misadjustment, and has significant influence on the learning (identification) performance. In practical use, the selection of step-size should guarantees the stability and convergence rate of the algorithm. To further improve the performance, the step-size η can be designed as a *variable* step-size. In Ref. [105], a self-adjusting step-size (SAS) was proposed to improve the performance of QIP criterion, i.e.,

$$\eta_k = \mu(\hat{V}_2(0) - \hat{V}_2(e)) \tag{4.69}$$

where $\hat{V}_2(0) = K_h(0)$ and $K_h(.)$ is a symmetric and unimodal kernel function (hence $\hat{V}_2(0) \geq \hat{V}_2(e)$).

The kernel width h is another important parameter that controls the smoothness of the performance surface. In general, the kernel width can be set manually or determined in advance by Silverman's rule. To make the algorithm converge to the global solution, one can start the algorithm with a large kernel width and decrease this parameter slowly during the course of adaptation, just like in stochastic annealing. In Ref. [185], an adaptive kernel width was proposed to improve the performance.

The BIG algorithm needs to acquire in advance all the training data, and hence is not suitable for online identification. To address this issue, one can use the sliding information gradient algorithm.

4.3.1.2 Sliding Information Gradient Algorithm

The sliding information gradient algorithm utilizes a set of recent error samples to estimate the error entropy. Specifically, the error samples used to calculate the error entropy at time k is as follows[9] :

$$\{e_{k-L+1}, \ldots, e_{k-1}, e_k\} \tag{4.70}$$

where L denotes the sliding data length $(L < N)$. Then the error entropy at time k can be estimated as

$$\hat{H}_{\phi,k}(e) = \frac{1}{L} \sum_{j=k-L+1}^{k} \psi \left(\frac{1}{L} \sum_{i=k-L+1}^{k} K_h(e_j - e_i) \right) \tag{4.71}$$

And hence, the sliding information gradient algorithm can be derived as

$$W_k = W_{k-1} - \eta \frac{\partial}{\partial W} \hat{H}_{\phi,k}(e) = W_{k-1} + \frac{\eta}{L^2} \sum_{j=k-L+1}^{k}$$
$$\left\{ \psi' \left(\frac{1}{L} \sum_{i=k-L+1}^{k} K_h(e_j - e_i) \right) \sum_{i=k-L+1}^{k} \left(K'_h(e_j - e_i) \left(\frac{\partial \hat{y}_j}{\partial W} - \frac{\partial \hat{y}_i}{\partial W} \right) \right) \right\} \tag{4.72}$$

---

[9] It should be noted that the error samples at time k are not necessarily to be a tap delayed sequence. A more general expression of the error samples can be $\{e_{1,k}, e_{2,k}, \ldots, e_{L,k}\}$.

For the case $\phi(x) = -x^2$ (corresponding to QIP), the above algorithm becomes

$$W_k = W_{k-1} - \frac{\eta}{L^2} \sum_{j=k-L+1}^{k} \sum_{i=k-L+1}^{k} \left(K'_h(e_j - e_i) \left(\frac{\partial \hat{y}_j}{\partial W} - \frac{\partial \hat{y}_i}{\partial W} \right) \right) \tag{4.73}$$

4.3.1.3 FRIG Algorithm

In the sliding information gradient algorithm, the error entropy can be estimated by a *forgetting recursive* method [64]. Assume at time $k-1$ the estimated error PDF is $\hat{p}_{k-1}(e)$, then the error PDF at time k can be estimated as

$$\hat{p}_k(e) = (1 - \lambda)\hat{p}_{k-1}(e) + \lambda K_h(e - e_k) \tag{4.74}$$

where λ is the forgetting factor. Therefore, we can calculate the empirical error entropy as follows:

$$\begin{aligned} \hat{H}_{\phi,k}(e) &= \frac{1}{L} \sum_{j=k-L+1}^{k} \psi(\hat{p}_k(e_j)) \\ &= \frac{1}{L} \sum_{j=k-L+1}^{k} \psi[(1 - \lambda)\hat{p}_{k-1}(e_j) + \lambda K_h(e_j - e_k)] \end{aligned} \tag{4.75}$$

If $\phi(x) = -x^2$ ($\psi(x) = -x$), there exists a recursive form [186]:

$$\hat{V}_{2,k}(e) = (1 - \lambda)\hat{V}_{2,k-1}(e) + \frac{\lambda}{L} \sum_{j=k-L+1}^{k} K_h(e_j - e_k) \tag{4.76}$$

where $\hat{V}_{2,k}(e)$ is the QIP at time k. Thus, we have the following algorithm:

$$\begin{aligned} W_k &= W_{k-1} + \eta \frac{\partial}{\partial W} \hat{V}_{2,k}(e) \\ &= W_{k-1} + \eta(1 - \lambda) \frac{\partial}{\partial W} \hat{V}_{2,k-1}(e) - \frac{\lambda\eta}{L} \sum_{j=k-L+1}^{k} K'_h(e_j - e_k) \left(\frac{\partial \hat{y}_j}{\partial W} - \frac{\partial \hat{y}_k}{\partial W} \right) \end{aligned}$$

$$\tag{4.77}$$

namely the FRIG algorithm. Compared with the sliding information gradient algorithm, the FRIG algorithm is computationally simpler and is more suitable for non-stationary system identification.

4.3.1.4   SIG Algorithm

In the empirical error entropy of (4.71), if dropping the outer averaging operator $(1/L\sum_{j=k-L+1}^{k}(.))$, one may obtain the *instantaneous* error entropy at time k:

$$\hat{H}_{\phi,k}(e) = \psi\left(\frac{1}{L}\sum_{i=k-L+1}^{k} K_h(e_k - e_i)\right) \tag{4.78}$$

The instantaneous error entropy is similar to the instantaneous error cost $\hat{R} = l(e_k)$, as both are obtained by removing the expectation operator (or averaging operator) from the original criterion function. The computational cost of the instantaneous error entropy (4.78) is $1/L$ of that of the empirical error entropy of Eq. (4.71). The gradient identification algorithm based on the instantaneous error entropy is called the SIG algorithm, which can be derived as

$$W_k = W_{k-1} - \eta\frac{\partial}{\partial W}\psi\left(\frac{1}{L}\sum_{i=k-L+1}^{k} K_h(e_k - e_i)\right)$$

$$= W_{k-1} + \frac{\eta}{L}\psi'\left(\frac{1}{L}\sum_{i=k-L+1}^{k} K_h(e_k - e_i)\right)\sum_{i=k-L+1}^{k}\left(K'_h(e_k - e_i)\left(\frac{\partial\hat{y}_k}{\partial W} - \frac{\partial\hat{y}_i}{\partial W}\right)\right) \tag{4.79}$$

If $\phi(x) = -x\log x$, we obtain the SIG algorithm under Shannon entropy criterion:

$$W_k = W_{k-1} - \frac{\eta}{L}\left(\frac{1}{L}\sum_{i=k-L+1}^{k} K_h(e_k-e_i)\right)^{-1}\sum_{i=k-L+1}^{k}\left(K'_h(e_k - e_i)\left(\frac{\partial\hat{y}_k}{\partial W} - \frac{\partial\hat{y}_i}{\partial W}\right)\right) \tag{4.80}$$

If $\phi(x) = -x^2$, we get the SIG algorithm under QIP criterion:

$$W_k = W_{k-1} - \frac{\eta}{L}\sum_{i=k-L+1}^{k}\left(K'_h(e_k - e_i)\left(\frac{\partial\hat{y}_k}{\partial W} - \frac{\partial\hat{y}_i}{\partial W}\right)\right) \tag{4.81}$$

The SIG algorithm (4.81) is actually the FRIG algorithm with $\lambda = 1$.

4.3.2   Parametric IG Algorithms

In IG algorithms described above, the error distribution is estimated by nonparametric KDE approach. With this approach, one is often confronted with the problem of how to choose a suitable value of the kernel width. An inappropriate choice of width will significantly deteriorate the performance of the algorithm. Though the effects of the kernel width on the shape of the performance surface and the eigenvalues of the Hessian at and around the optimal solution have been carefully investigated [102], at present the choice of the kernel width is still a difficult task. Thus, a certain parameterized density estimation, which does not involve the choice of kernel width, sometimes might be more practical. Especially, if some prior knowledge about the data distribution is available, the parameterized density estimation may achieve a better accuracy than nonparametric alternatives.

Next, we discuss the parametric IG algorithms that adopt parametric approaches to estimate the error distribution. To simplify the discussion, we only present the parametric SIG algorithm.

In general, the SIG algorithm can be expressed as

$$W_k = W_{k-1} - \eta \frac{\partial}{\partial W} \psi(\hat{p}(e_k)) \tag{4.82}$$

where $\hat{p}(e_k)$ is the value of the error PDF at e_k estimated based on the error samples $\{e_{k-L+1}, \ldots, e_{k-1}, e_k\}$. By KDE approach, we have

$$\hat{p}(e_k) = \frac{1}{L} \sum_{i=k-L+1}^{k} K_h(e_k - e_i) \tag{4.83}$$

Now we use a parametric approach to estimate $\hat{p}(e_k)$. Let's consider the exponential (maximum entropy) PDF form:

$$p(e) = \exp\left(-\lambda_0 - \sum_{r=1}^{K} \lambda_r g_r(e) \right) \tag{4.84}$$

where the parameters λ_r $(r = 0, 1, \ldots, K)$ can be estimated by some classical estimation methods like the ML estimation. After obtaining the estimated parameter values $\hat{\lambda}_r$, one can calculate $\hat{p}(e_k)$ as

$$\hat{p}(e_k) = \exp\left(-\hat{\lambda}_0 - \sum_{r=1}^{K} \hat{\lambda}_r g_r(e_k) \right) \tag{4.85}$$

Substituting Eq. (4.85) into Eq. (4.82), we obtain the following parametric SIG algorithm:

$$W_k = W_{k-1} - \eta \frac{\partial}{\partial W} \psi\left(\exp\left(-\hat{\lambda}_0 - \sum_{r=1}^{K} \hat{\lambda}_r g_r(e_k) \right) \right) \tag{4.86}$$

If adopting Shannon entropy ($\psi(x) = -\log(x)$), the algorithm becomes

$$
\begin{aligned}
W_k &= W_{k-1} + \eta \frac{\partial}{\partial W} \log\left(\exp\left(-\hat{\lambda}_0 - \sum_{r=1}^{K} \hat{\lambda}_r g_r(e_k) \right) \right) \\
&= W_{k-1} + \eta \frac{\partial}{\partial W} \left(-\hat{\lambda}_0 - \sum_{r=1}^{K} \hat{\lambda}_r g_r(e_k) \right) \\
&= W_{k-1} - \eta \sum_{r=1}^{K} \hat{\lambda}_r g'_r(e_k) \frac{\partial e_k}{\partial W} \\
&= W_{k-1} + \eta \sum_{r=1}^{K} \hat{\lambda}_r g'_r(e_k) \frac{\partial \hat{y}_k}{\partial W}
\end{aligned}
\tag{4.87}
$$

The selection of the PDF form is very important. Some typical PDF forms are as follows [187–190]:

1. $p(e) = \exp(-\lambda_0 - \lambda_1 e - \lambda_2 e^2 - \lambda_3 \log(1 + e^2) - \lambda_4 \sin(e) - \lambda_5 \cos(e))$
2. $p(e) = \exp(-\lambda_0 - \lambda_1 e - \lambda_2 e^2 - \lambda_3 \log(1 + e^2))$
3. Generalized Gaussian density (GGD) model [191]:

$$
p(e) = \frac{\alpha}{2\beta\Gamma(1/\alpha)} \exp(-(|e - \mu|/\beta)^\alpha)
\tag{4.88}
$$

where $\beta = \sigma\sqrt{\Gamma(1/\alpha)/\Gamma(3/\alpha)}$, $\sigma > 0$ is the standard deviation, and $\Gamma(.)$ is the Gamma function:

$$
\Gamma(z) = \int_0^\infty x^{z-1} e^{-x} \, dx, \quad z > 0
\tag{4.89}
$$

In the following, we present the SIG algorithm based on GGD model.

The GGD model has three parameters: location (mean) parameter μ, shape parameter α, and dispersion parameter β. It has simple yet flexible functional forms and could approximate a large number of statistical distributions, and is widely used in image coding, speech recognition, and BSS, etc. The GGD densities include Gaussian ($\alpha = 2$) and Laplace ($\alpha = 1$) distributions as special cases. Figure 4.4 shows the GGD distributions for several shape parameters with zero mean and deviation 1.0. It is evident that smaller values of the shape parameter correspond to heavier tails and therefore to sharper distributions. In the limiting cases, as $\alpha \to \infty$, the GGD becomes close to the uniform distribution, whereas as $\alpha \to 0+$, it approaches an impulse function (δ-distribution).

Utilizing the GGD model to estimate the error distribution is actually to estimate the parameters μ, α, and β based on the error samples. Up to now, there are many methods on how to estimate the GGD parameters [192]. Here, we only discuss the moment matching method (method of moments).

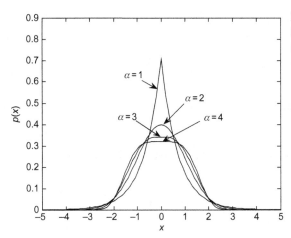

Figure 4.4 GGD distributions for different shape parameters α ($\mu = 0$, $\sigma = 1.0$).

The order-r absolute central moment of GGD distribution can be calculated as

$$m_r = E[|X - \mu|^r] = \int_{-\infty}^{+\infty} |x - \mu|^r p(x)\mathrm{d}x = \beta^r \Gamma((r + 1)/\alpha)/\Gamma(1/\alpha) \tag{4.90}$$

Hence we have

$$\frac{m_r}{\sqrt{m_{2r}}} = \frac{\Gamma((r + 1)/\alpha)}{\sqrt{\Gamma((2r + 1)/\alpha)\Gamma(1/\alpha)}} \tag{4.91}$$

The right-hand side of Eq. (4.91) is a function of α, denoted by $R_r(\alpha)$. Thus the parameter α can be expressed as

$$\alpha = R_r^{-1}\left(m_r/\sqrt{m_{2r}}\right) \tag{4.92}$$

where $R_r^{-1}(.)$ is the inverse of function $R_r(.)$. Figure 4.5 shows the curves of the inverse function $y = R_r^{-1}(x)$ when $r = 1, 2$.

According to Eq. (4.92), based on the moment matching method one can estimate the parameters μ, α, and β as follows:

$$\begin{cases} \hat{\mu}_k = \dfrac{1}{L} \displaystyle\sum_{i=k-L+1}^{k} e_i \\[2ex] \hat{\alpha}_k = R_r^{-1}\left(\left[\dfrac{1}{L} \displaystyle\sum_{i=k-L+1}^{k} |e_i - \hat{\mu}_k|^r\right] \middle/ \sqrt{\dfrac{1}{L} \displaystyle\sum_{i=k-L+1}^{k} (e_i - \hat{\mu}_k)^{2r}}\right) \\[2ex] \hat{\beta}_k = \sqrt{\dfrac{1}{L} \displaystyle\sum_{i=k-L+1}^{k} (e_i - \hat{\mu}_k)^2} \times \sqrt{\dfrac{\Gamma(1/\hat{\alpha}_k)}{\Gamma(3/\hat{\alpha}_k)}} \end{cases} \tag{4.93}$$

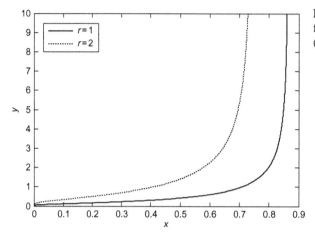

Figure 4.5 Inverse functions $y = R_r^{-1}(x)$ ($r = 1, 2$).

where the subscript k represents that the values are estimated based on the error samples $\{e_{k-L+1}, \ldots, e_{k-1}, e_k\}$. Thus $\hat{p}(e_k)$ will be

$$\hat{p}(e_k) = \frac{\hat{\alpha}_k}{2\hat{\beta}_k \Gamma(1/\hat{\alpha}_k)} \exp\left(-\left(\frac{|e_k - \hat{\mu}_k|}{\hat{\beta}_k}\right)^{\hat{\alpha}_k}\right) \tag{4.94}$$

Substituting Eq. (4.94) into Eq. (4.82) and letting $\psi(x) = -\log x$ (Shannon entropy), we obtain the SIG algorithm based on GGD model:

$$\begin{aligned}
W_k &= W_{k-1} - \eta \frac{\partial}{\partial W}\left\{-\log \hat{p}(e_k)\right\} \\
&= W_{k-1} + \eta \frac{\hat{\alpha}_k |e_k - \hat{\mu}_k|^{\hat{\alpha}_k - 1} \text{sign}(e_k - \hat{\mu}_k)}{\hat{\beta}_k^{\hat{\alpha}_k}} \frac{\partial \hat{y}_k}{\partial W}
\end{aligned} \tag{4.95}$$

where $\hat{\mu}_k$, $\hat{\alpha}_k$, and $\hat{\beta}_k$ are calculated by Eq. (4.93).

To make a distinction, we denote "SIG-kernel" and "SIG-GGD" the SIG algorithms based on kernel approach and GGD densities, respectively. Compared with the SIG-kernel algorithm, the SIG-GGD just needs to estimate the three parameters of GGD density, without resorting to the choice of kernel width and the calculation of kernel function.

Comparing Eqs. (4.95) and (4.28), we find that when $\hat{\mu}_k \approx 0$, the SIG-GGD algorithm can be considered as an LMP algorithm with adaptive order $\hat{\alpha}_k$ and variable step-size $\eta/\hat{\beta}_k^{\hat{\alpha}_k}$. In fact, under certain conditions, the SIG-GGD algorithm will converge to a certain LMP algorithm with fixed order and step-size. Consider the FIR system identification, in which the plant and the adaptive model are both FIR filters with the same order, and the additive noise $\{n_k\}$ is zero mean, ergodic, and stationary. When the model weight vector W_k converges to the neighborhood of the plant weight vector W^*, we have

$$e_k = z_k - \hat{y}_k = \tilde{W}_k^T X_k + n_k \approx n_k \tag{4.96}$$

where $\tilde{W}_k = W^* - W_k$ is the *weight error vector*. In this case, the estimated values of the parameters in SIG-GGD algorithm will be

$$\begin{cases} \hat{\mu}_k \approx \dfrac{1}{L} \displaystyle\sum_{i=k-L+1}^{k} n_i \approx 0 \\[2ex] \hat{\alpha}_k \approx R_r^{-1} \left(\left[\dfrac{1}{L} \displaystyle\sum_{i=k-L+1}^{k} |n_i - \hat{\mu}_k|^r \right] \Big/ \sqrt{\dfrac{1}{L} \displaystyle\sum_{i=k-L+1}^{k} (n_i - \hat{\mu}_k)^{2r}} \right) \\[2ex] \hat{\beta}_k \approx \sqrt{\dfrac{1}{L} \displaystyle\sum_{i=k-L+1}^{k} (n_i - \hat{\mu}_k)^2} \times \sqrt{\dfrac{\Gamma(1/\hat{\alpha}_k)}{\Gamma(3/\hat{\alpha}_k)}} \end{cases} \tag{4.97}$$

Since noise $\{n_k\}$ is an ergodic and stationary process, if L is large enough, the estimated values of the three parameters will tend to some constants, and consequently, the SIG-GGD algorithm will converge to a certain LMP algorithm with fixed order and step-size. Clearly, if noise n_k is Gaussian distributed, the SIG-GGD algorithm will converge to the LMS algorithm ($\hat{\alpha}_k \approx 2$), and if n_k is Laplacian distributed, the algorithm will converge to the LAD algorithm ($\hat{\alpha}_k \approx 1$). In Ref. [28], it has been shown that under slow adaptation, the LMS and LAD algorithms are, respectively, the optimum algorithms for the Gaussian and Laplace interfering noises. We may therefore conclude that the SIG-GGD algorithm has the ability to adjust its parameters so as to automatically switch to a certain optimum algorithm.

There are two points that deserve special mention concerning the implementation of the SIG-GGD algorithm: (i) since there is no analytical expression, the calculation of the inverse function $R_r^{-1}(.)$ needs to use look-up table or some interpolation method and (ii) in order to avoid too large gradient and ensure the stability of the algorithm, it is necessary to set an upper bound on the parameter $\hat{\alpha}_k$.

4.3.3 Fixed-Point Minimum Error Entropy Algorithm

Given a mapping $f : A \to A$, the fixed points are solutions of iterative equation $x = f(x)$, $x \in A$. The fixed-point (FP) iteration is a numerical method of computing fixed points of iterated functions. Given an initial point $x_0 \in A$, the FP iteration algorithm is

$$x_{k+1} = f(x_k), k = 0, 1, 2, \ldots \tag{4.98}$$

where k is the iterative index. If f is a function defined on the real line with real values, and is Lipschitz continuous with Lipschitz constant smaller than 1.0, then f

has precisely one fixed point, and the FP iteration converges toward that fixed point for any initial guess x_0. This result can be generalized to any metric space.

The FP algorithm can be applied in parameter identification under MEE criterion [64]. Let's consider the QIP criterion:

$$\hat{V}_2(e) = \frac{1}{N^2} \sum_{j=1}^{N} \sum_{i=1}^{N} K_h(e_j - e_i) \tag{4.99}$$

under which the optimal parameter (weight vector) W^* satisfies

$$\frac{\partial}{\partial W} \hat{V}_2(e) = -\frac{1}{N^2} \sum_{j=1}^{N} \sum_{i=1}^{N} K'_h(e_j - e_i) \left(\frac{\partial \hat{y}_j}{\partial W} - \frac{\partial \hat{y}_i}{\partial W} \right) \bigg|_{W=W^*} = 0 \tag{4.100}$$

If the model is an FIR filter, and the kernel function is the Gaussian function, we have

$$\frac{1}{N^2 h^2} \sum_{j=1}^{N} \sum_{i=1}^{N} (e_j - e_i) K_h(e_j - e_i)(X_j - X_i) = 0 \tag{4.101}$$

One can write Eq. (4.101) in an FP iterative form (utilizing $e_k = z_k - \hat{y}_k = z_k - W_k^T X_k$):

$$W^* = f(W^*)$$

$$= \left[\frac{1}{N^2} \sum_{j=1}^{N} \sum_{i=1}^{N} K_h(e_j - e_i)(X_j - X_i)(X_j - X_i)^T \right]^{-1}$$

$$\times \left[\frac{1}{N^2} \sum_{j=1}^{N} \sum_{i=1}^{N} K_h(e_j - e_i)(z_j - z_i)(X_j - X_i) \right] \tag{4.102}$$

Then we have the following FP algorithm:

$$W_{k+1} = f(W_k) = R_E(W_k)^{-1} P_E(W_k) \tag{4.103}$$

where

$$\begin{cases} R_E(W_k) = \frac{1}{N^2} \sum_{j=1}^{N} \sum_{i=1}^{N} K_h(e_j - e_i)(X_j - X_i)(X_j - X_i)^T \\ P_E(W_k) = \frac{1}{N^2} \sum_{j=1}^{N} \sum_{i=1}^{N} K_h(e_j - e_i)(z_j - z_i)(X_j - X_i) \end{cases} \tag{4.104}$$

The above algorithm is called the fixed-point minimum error entropy (FP-MEE) algorithm. The FP-MEE algorithm can also be implemented by using the forgetting recursive form [194], i.e.,

$$W_{k+1} = \overline{f}(W_k) = \overline{R}_E(W_k)^{-1}\overline{P}_E(W_k) \tag{4.105}$$

where

$$
\begin{cases}
\overline{R}_E(W_{k+1}) = \lambda\overline{R}_E(W_k) + \dfrac{1-\lambda}{L}\displaystyle\sum_{i=k-L+1}^{k}K_h(e_{k+1}-e_i)(X_{k+1}-X_i)(X_{k+1}-X_i)^T \\[4mm]
\overline{P}_E(W_{k+1}) = \lambda\overline{P}_E(W_k) + \dfrac{1-\lambda}{L}\displaystyle\sum_{i=k-L+1}^{k}K_h(e_{k+1}-e_i)(z_{k+1}-z_i)(X_{k+1}-X_i)
\end{cases}
$$

$$\tag{4.106}$$

This is the recursive fixed-point minimum error entropy (RFP-MEE) algorithm.

In addition to the parameter search algorithms described above, there are many other parameter search algorithms to minimize the error entropy. Several advanced parameter search algorithms are presented in Ref. [104], including the conjugate gradient (CG) algorithm, Levenberg−Marquardt (LM) algorithm, quasi-Newton method, and others.

4.3.4 Kernel Minimum Error Entropy Algorithm

System identification algorithms under MEE criterion can also be derived in kernel space. Existing KAF algorithms are mainly based on the MSE (or least squares) criterion. MSE is not always a suitable criterion especially in nonlinear and non-Gaussian situations. Hence, it is attractive to develop a new KAF algorithm based on a non-MSE (nonquadratic) criterion. In Ref. [139], a KAF algorithm under the maximum correntropy criterion (MCC), namely the kernel maximum correntropy (KMC) algorithm, has been developed. Similar to the KLMS, the KMC is also a stochastic gradient algorithm in RKHS. If the kernel function used in correntropy, denoted by κ_c, is the Gaussian kernel, the KMC algorithm can be derived as [139]

$$\Omega_k = \Omega_{k-1} + \eta\kappa_c(e_k)e_k\varphi_k \tag{4.107}$$

where Ω_k denotes the estimated weight vector at iteration k in a high-dimensional feature space $\mathbb{F}_\kappa$ induced by Mercer kernel κ and φ_k is a feature vector obtained by transforming the input vector X_k into the feature space through a nonlinear mapping φ. The KMC algorithm can be regarded as a KLMS algorithm with variable step-size $\eta\kappa_c(e_k)$.

In the following, we will derive a KAF algorithm under the MEE criterion. Since the ϕ-entropy is a very general and flexible entropy definition, we use the ϕ-entropy of the error as the adaptation criterion. In addition, for simplicity we adopt the instantaneous error entropy (4.78) as the cost function. Then, one can easily derive the following kernel minimum error entropy (KMEE) algorithm:

$$\Omega_k = \Omega_{k-1} - \eta \frac{\partial}{\partial \Omega} (\hat{H}_{\phi,k}(e))$$

$$= \Omega_{k-1} - \eta \frac{\partial}{\partial \Omega} \left[\psi \left(\frac{1}{L} \sum_{i=k-L+1}^{k} K_h(e_k - e_i) \right) \right]$$

$$= \Omega_{k-1} + \frac{\eta}{L} \psi' \left(\frac{1}{L} \sum_{i=k-L+1}^{k} K_h(e_k - e_i) \right) \sum_{i=k-L+1}^{k} \left(K'_h(e_k - e_i) \left(\frac{\partial \hat{y}_k}{\partial \Omega} - \frac{\partial \hat{y}_i}{\partial \Omega} \right) \right)$$

$$= \Omega_{k-1} + \frac{\eta}{L} \psi' \left(\frac{1}{L} \sum_{i=k-L+1}^{k} K_h(e_k - e_i) \right) \sum_{i=k-L+1}^{k} (K'_h(e_k - e_i)(\varphi_k - \varphi_i)) \tag{4.108}$$

The KMEE algorithm (4.108) is actually the SIG algorithm in kernel space. By selecting a certain ϕ function, we can obtain a specific KMEE algorithm. For example, if setting $\phi(x) = -x \log x$ (i.e., $\psi(x) = -\log x$), we get the KMEE under Shannon entropy criterion:

$$\Omega_k = \Omega_{k-1} - \eta \frac{\sum_{i=k-L+1}^{k} \left(K'_h(e_k - e_i)(\varphi_k - \varphi_i) \right)}{\sum_{i=k-L+1}^{k} K_h(e_k - e_i)} \tag{4.109}$$

The weight update equation of Eq. (4.108) can be written in a compact form:

$$\Omega_k = \Omega_{k-1} + \eta \Phi_k h_\phi(e_k) \tag{4.110}$$

where $e_k = [e_{k-L+1}, e_{k-L+2}, \ldots, e_k]^T$, $\Phi_k = [\varphi_{k-L+1}, \varphi_{k-L+2}, \ldots, \varphi_k]$, and $h_\phi(e_k)$ is a vector-valued function of e_k, expressed as

$$h_\phi(e_k) = \frac{1}{L} \psi' \left(\frac{1}{L} \sum_{i=k-L+1}^{k} K_h(e_k - e_i) \right) \times \begin{pmatrix} -K'_h(e_k - e_{k-L+1}) \\ \vdots \\ -K'_h(e_k - e_{k-1}) \\ \sum_{i=k-L+1}^{k-1} K'_h(e_k - e_i) \end{pmatrix} \tag{4.111}$$

The KMEE algorithm is similar to the KAPA [178], except that the error vector e_k in KMEE is nonlinearly transformed by the function $h_\phi(.)$. The learning rule of the KMEE in the original input space can be written as ($f_0 = 0$)

$$f_k = f_{k-1} + \eta \mathbf{K}_k h_\phi(e_k) \tag{4.112}$$

where $\mathbf{K}_k = [\kappa(X_{k-L+1}, .), \kappa(X_{k-L+2}, .), \ldots, \kappa(X_k, .)]$.

The learned model by KMEE has the same structure as that learned by KLMS, and can be represented as a linear combination of kernels centered in each data points:

$$f_k(.) = \sum_{j=1}^{k} \alpha_j \kappa(X_j, .) \tag{4.113}$$

where the coefficients (at iteration k) are updated as follows:

$$\alpha_j = \begin{cases} \dfrac{\eta}{L} \psi' \left(\dfrac{1}{L} \sum_{i=k-L+1}^{k} K_h(e_k - e_i) \right) \sum_{i=k-L+1}^{k-1} K'_h(e_k - e_i), & j = k \\ \alpha_j - \dfrac{\eta}{L} \psi' \left(\dfrac{1}{L} \sum_{i=k-L+1}^{k} K_h(e_k - e_i) \right) K'_h(e_k - e_j), & k - L < j < k \\ \alpha_j, & 1 \le j \le k - L \end{cases} \tag{4.114}$$

The pseudocode for KMEE is summarized in Table 4.2.

4.3.5 Simulation Examples

In the following, we present several simulation examples to demonstrate the performance (accuracy, robustness, convergence rate, etc.) of the identification algorithms under MEE criterion.

Table 4.2 The KMEE Algorithm

Initialization:
a. Assigning the ϕ function and the kernel functions κ and K_h;
b. Choose the step-size η, and the sliding data length L;
c. Initialize the center set $C = \{X_1\}$, and the coefficient vector $\alpha = [\eta z_1]^T$;
d. Initialize the window of L errors: $e = \{0, \ldots, 0\}$.
Computation:
while$\{X_k, z_k\}$ $(k > 1)$ available **do**

1 Allocate a new unit: $C = \{C, X_k\}$, $a - [a^T, 0]^T$
2. Update the window of errors:-

$$\begin{cases} e(i) = e(i + 1), & \text{for} \quad i = 1, \ldots, L - 1 \\ e(L) = e_k \end{cases}$$

where $e_k = z_k - \sum_{j=1}^{k-1} \alpha_j \kappa(X_j, X_k)$

3. Update the coefficient vector $a = [\alpha_1, \ldots, \alpha_k]^T$ using (4.114).
end while

Example 4.1 [102] Assume that both the unknown system and the model are two-dimensional ADALINEs, i.e.,

$$
\begin{cases}
z_k = w_1^* x_{1,k} + w_2^* x_{2,k} + n_k \\
\hat{y}_k = w_1 x_{1,k} + w_2 x_{2,k}
\end{cases}
\tag{4.115}
$$

where unknown weight vector is $W^* = [w_1^*, w_2^*]^T = [1.0, 2.0]^T$, and n_k is the independent and zero-mean Gaussian noise. The goal is to identify the model parameters $W = [w_1, w_2]^T$ under noises of different signal-to-noise ratios (SNRs). For each noise energy, 100 independent Monte-Carlo simulations are performed with N ($N = 10, 20, 50, 100$) training data that are chosen randomly. In the simulation, the BIG algorithm under QIP criterion is used, and the kernel function K_h is the Gaussian function with bandwidth $h = 1.0$. Figure 4.6 shows the average distance between actual and estimated optimal weight vector. For comparison purpose, the figure also includes the identification results (by solving the Wiener-Hopf equations) under MSE criterion. Simulation results indicate that, when SNR is higher (SNR > 10 *dB*), the MEE criterion achieves much better accuracy than the MSE criterion (or requires less training data when achieving the same accuracy).

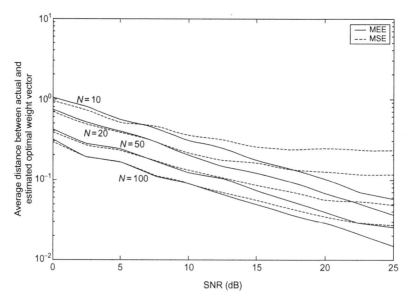

Figure 4.6 Comparison of the performance of MEE against MSE (adopted from Ref. [102]).

Example 4.2 [100] Identify the following nonlinear dynamic system:

$$
\begin{cases}
x_{1,k+1} = \left(\dfrac{x_{1,k}}{1+x_{1,k}^2} + 1\right).\sin x_{2,k} \\[3mm]
x_{2,k+1} = x_{2,k}.\cos x_{2,k} + \exp\left(-\dfrac{x_{1,k}^2 + x_{2,k}^2}{8}\right) + \dfrac{u_k^3}{1+u_k^2 + 0.5\cos(x_{1,k}+x_{2,k})} \\[3mm]
y_k = \dfrac{x_{1,k}}{1+0.5\sin x_{2,k}} + \dfrac{x_{2,k}}{1+0.5\sin x_{1,k}}
\end{cases}
$$

$$(4.116)$$

where $x_{1,k}$ and $x_{2,k}$ are the state variables and u_k is the input signal. The identification model is the time delay neural network (TDNN), where the network structure is an MLP with multi-input, single hidden layer, and single output. The input vector of the neural network contains the current input and output and their past values of the nonlinear system, that is, the training data can be expressed as

$$
\{[u_k, u_{k-1}, \ldots, u_{k-n_u}, y_{k-1}, \ldots, y_{k-n_y}]^T, y_k\}
$$

$$(4.117)$$

In this example, n_u and n_y are set as $n_u = n_y = 6$. The number of hidden units is set at 7, and the symmetric sigmoid function is selected as the activation function. In addition, the number of training data is $N = 100$. We continue to compare the performance of MEE (using the BIG algorithm under QIP criterion) to MSE. For each criterion, the TDNN is trained starting from 50 different initial weight vectors, and the best solution (the one with the highest QIP or lowest MSE) among the 50 candidates is selected to test the performance.[10] Figure 4.7 illustrates the probability densities of the error between system actual output and TDNN output with 10,000 testing data. One can see that the MEE criterion achieves a higher peak around the zero error. Figure 4.8 shows the probability densities of system actual output (desired output) and model output. Evidently, the output of the model trained under MEE criterion matches the desired output better.

Example 4.3 [190] Compare the performances of SIG-kernel, SIG-GGD, and LMP family algorithms (LAD, LMS, LMF, etc.). Assume that both the unknown system and the model are FIR filters:

$$
\begin{cases}
G^*(z) = 0.1 + 0.3z^{-1} + 0.5z^{-2} + 0.3z^{-3} + 0.1z^{-4} \\
G(z) = w_0 + w_1 z^{-1} + w_2 z^{-2} + w_3 z^{-3} + w_4 z^{-4}
\end{cases}
$$

$$(4.118)$$

---

[10] Since error entropy is shift-invariant, after training under MEE criterion the bias value of the output PE was adjusted so as to yield zero-mean error over the training set.

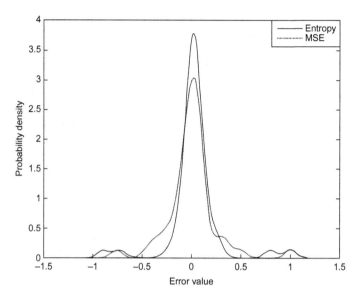

Figure 4.7 Probability densities of the error between system actual output and model output (adopted from Ref. [100]).

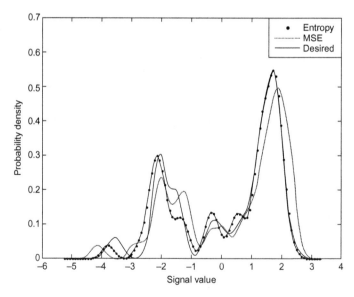

Figure 4.8 Probability densities of system actual output and model output (adopted from Ref. [100]).

where $G^*(z)$ and $G(z)$ denote the transfer functions of the system and the model, respectively. The initial weight vector of the model is set to be $W_0 = [0, 0, 0, 0, 0]^T$, and the input signal is white Gaussian noise with zero mean and unit power (variance). The kernel function in SIG-kernel algorithm is the Gaussian kernel with bandwidth determined by Silverman rule. In SIG-GGD algorithm, we set $r = 2$, and to avoid large gradient, we set the upper bound of α_k at 4.0.

In the simulation, we consider four noise distributions (Laplace, Gaussian, Uniform, MixNorm), as shown in Figure 4.9. For each noise distribution, the average convergence curves, over 100 independent Monte Carlo simulations, are illustrated in Figure 4.10, where WEP denotes the weight error power, defined as

$$\text{WEP} \triangleq E\left[\left\|\tilde{W}_k\right\|^2\right] = E[\tilde{W}_k^T \tilde{W}_k] \tag{4.119}$$

where $\tilde{W}_k = W^* - W_k$ is the weight error vector (the difference between desired and estimated weight vectors) and $\left\|\tilde{W}_k\right\|$ is the weight error norm. Table 4.3 lists the average identification results (mean $\pm$ deviation) of $w_2(w_2^* = 0.5)$. Further, the average evolution curves of α_k in SIG-GGD are shown in Figure 4.11.

From the simulation results, we have the following observations:

i. The performances of LAD, LMS, and LMF depend crucially on the distribution of the disturbance noise. These algorithms may achieve the smallest misadjustment for a certain noise distribution (e.g., the LMF performs best in uniform noise); however, for other noise distributions, their performances may deteriorate dramatically (e.g., the LMF performs worst in Laplace noise).

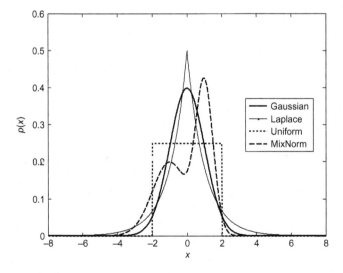

Figure 4.9 Four PDFs of the additive noise (adopted from Ref. [190]).

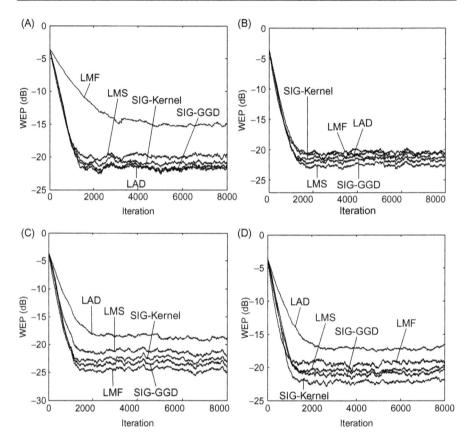

Figure 4.10 Average convergence curves of several algorithms for different noise distributions: (A) Laplace, (B) Gaussian, (C) Uniform, and (D) MixNorm (adopted from Ref. [190]).

ii. Both SIG-GGD and SIG-Kernel are robust to noise distribution. In the case of symmetric and unimodal noises (e.g., Laplace, Gaussian, Uniform), SIG-GGD may achieve a smaller misadjustment than SIG-Kernel. Though in the case of nonsymmetric and nonunimodal noises (e.g., MixNorm), SIG-GGD may be not as good as SIG-Kernel, it is still better than most of the LMP algorithms.

iii. Near the convergence, the parameter α_k in SIG-GGD converges approximately to 1, 2, and 4 (note that α_k is restricted to $\alpha_k \leq 4$ artificially) when disturbed by, respectively, Laplace, Gaussian, and Uniform noises. This confirms the fact that SIG-GGD has the ability to adjust its parameters so as to switch to a certain optimum algorithm.

Example 4.4 [194] Apply the RFP-MEE algorithm to identify the following FIR filter:

$$G^*(z) = 0.1 + 0.2z^{-1} + 0.3z^{-2} + 0.4z^{-3} + 0.5z^{-4} \\ + 0.4z^{-5} + 0.3z^{-6} + 0.2z^{-7} + 0.1z^{-8} \tag{4.120}$$

Table 4.3 Average Identification Results of w_2 Over 100 Monte Carlo Simulations

	SIG-Kernel	SIG-GGD	LAD	LMS	LMF
Laplace	0.5034 ± 0.0419	0.5019 ± 0.0366	0.5011 ± 0.0352	0.5043 ± 0.0477	0.5066 ± 0.0763
Gaussian	0.5035 ± 0.0375	0.5026 ± 0.0346	0.5038 ± 0.0402	0.5020 ± 0.0318	0.5055 ± 0.0426
Uniform	0.4979 ± 0.0323	0.4995 ± 0.0311	0.5064 ± 0.0476	0.5035 ± 0.0357	0.4999 ± 0.0263
MixGaussian	0.4997 ± 0.0356	0.5014 ± 0.0449	0.4972 ± 0.0576	0.5021 ± 0.0463	0.5031 ± 0.0523

(adopted from Ref. [190])

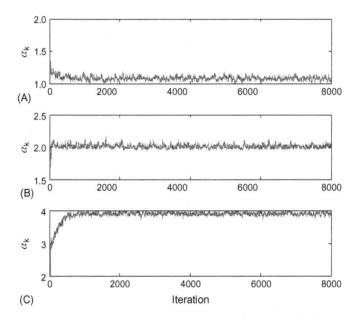

(A)

(B)

(C)

Iteration

Figure 4.11 Evolution curves of α_k over 100 Monte Carlo runs: (A) Laplace, (B) Gaussian, and (C) Uniform
(adopted from Ref. [190]).

The adaptive model is also an FIR filter with equal order. The input to both the plant and adaptive model is white Gaussian noise with unit power. The observation noise is white Gaussian distributed with zero mean and variance 10^{-10}. The main objective is to investigate the effect of the forgetting factor on the convergence speed and convergence accuracy (WEP after convergence) of the RFP-MEE algorithm. Figure 4.12 shows the convergence curves of the RFP-MEE with different forgetting factors. One can see that smaller forgetting factors result in faster convergence speed and larger steady-state WEP. This result conforms to the well-known general behavior of the forgetting factor in recursive estimates. Thus, selecting a proper forgetting factor for RFP-MEE must consider the intrinsic trade-off between convergence speed and identification accuracy.

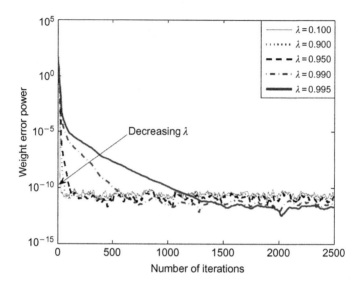

Figure 4.12 Convergence curves of RFP-MEE algorithm with different forgetting factors (adopted from Ref. [194]).

Example 4.5 This example aims to demonstrate the performance of KMEE (with Shannon entropy or QIP criterion). For comparison purpose, we also show the performances of several other KAF algorithms: KLMS, KMC, and KAPA. Let's consider the nonlinear system identification, where the nonlinear system is as follows [195]:

$$
\begin{aligned}
y_k = {}& (0.8 - 0.5 \exp(- y_{k-1}^2))y_{k-1} \\
& - (0.3 + 0.9\exp(- y_{k-1}^2))y_{k-2} \\
& + 0.1 \sin(3.1415926 y_{k-1}) + n_k
\end{aligned}
\tag{4.121}
$$

The noise n_k is of symmetric α-stable ($S\alpha S$) distribution with characteristic function

$$
\psi(\omega) = \exp(- \gamma|\omega|^\alpha)
\tag{4.122}
$$

where $\gamma = 0.005$, $0 < \alpha \le 2.0$. When $\alpha = 2.0$, the distribution is a zero-mean Gaussian distribution with variance 0.01; while when $\alpha < 2.0$, the distribution corresponds to an impulsive noise with infinite variance.

Figure 4.13 illustrates the average learning curves (over 200 Monte Carlo runs) for different α values and Table 4.4 lists the testing MSE at final iteration. In the simulation, 1000 samples are used for training and another 100 clean samples are used for testing (the filter is fixed during the testing phase). Further, all the kernels (kernel of RKHS, kernel of correntropy, and kernel for density estimation) are selected to be the Gaussian kernel. The kernel parameter for RKHS is set at 0.2,

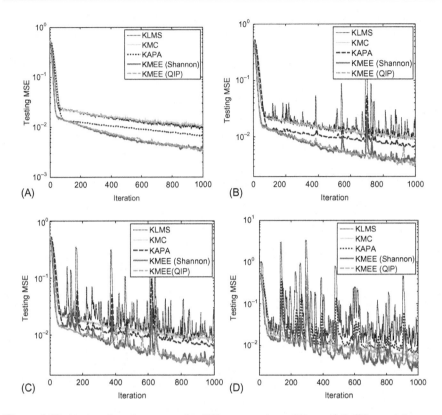

Figure 4.13 Average learning curves for different α values: (A) $\alpha = 2.0$, (B) $\alpha = 1.9$, (C) $\alpha = 1.8$, (D) $\alpha = 1.5$.

Table 4.4 Testing MSE at Final Iteration for Different α Values

Algorithms	Testing MSE			
	$\alpha = 2.0$	$\alpha = 1.9$	$\alpha = 1.8$	$\alpha = 1.5$
KLMS	0.0095 ± 0.0079	0.0134 ± 0.0647	0.0136 ± 0.1218	0.0203 ± 0.3829
KMC	0.0103 ± 0.0082	0.0096 ± 0.0089	0.0088 ± 0.0086	0.0063 ± 0.0064
KAPA	0.0067 ± 0.0015	0.0069 ± 0.0055	0.0073 ± 0.0078	0.0072 ± 0.0205
KMEE (Shannon)	$\mathbf{0.0040 \pm 0.0027}$	$\mathbf{0.0035 \pm 0.0028}$	$\mathbf{0.0035 \pm 0.0051}$	$\mathbf{0.0041 \pm 0.0180}$
KMEE (QIP)	$\mathbf{0.0035 \pm 0.0020}$	$\mathbf{0.0034 \pm 0.0022}$	$\mathbf{0.0036 \pm 0.0046}$	$\mathbf{0.0048 \pm 0.0138}$

kernel size for correntropy is 0.4, and kernel bandwidth for density estimation is 1.0. The sliding data lengths for KMEE and KAPA are both set at $L = 10$. The step-sizes for KLMS, KMC, KAPA, KMEE (Shannon), and KMEE (QIP), are, respectively, set at 0.8, 1.0, 0.05, 1.0, and 2.0. These parameters are experimentally selected to achieve the desirable performance. From Figure 4.13 and Table 4.4, one can see that the KMEE algorithms outperform all other algorithms except for the case of small α, the KMC algorithm may achieve a relatively smaller deviation in testing MSE. Simulation results also show that the performances of KMEE (Shannon) and KMEE (QIP) are very close.

4.4 Convergence Analysis

Next, we analyze the convergence properties of the parameter identification under MEE criterion. For simplicity, we consider only the ADALINE model (which includes the FIR filter as a special case). The convergence analysis of error entropy minimization algorithms is, in general, rather complicated. This is mostly because

1. the objective function (the empirical error entropy) is not only related to the current error but also concerned with the past error values;
2. the entropy function and the kernel function are nonlinear functions;
3. the shape of performance surface is very complex (nonquadratic and nonconvex).

There are two ways for the convergence analysis of such algorithms: (i) using the Taylor series expansion near the optimal solution to obtain an approximate linearization algorithm, and then performing the convergence analysis for the linearization algorithm and (ii) applying the energy conservation relation to analyze the convergence behavior [106]. The first approach is relatively simple, but only the approximate analysis results near the optimal solution can be achieved. With the second approach it is possible to acquire rigorous analysis results of the convergence, but usually more assumptions are needed. In the following, we first briefly introduce the first analysis approach, and then focus on the second approach, performing the mean square convergence analysis based on the energy conservation relation. The following analysis is mainly aimed at the nonparametric IG algorithms with KDE.

4.4.1 Convergence Analysis Based on Approximate Linearization

In Ref. [102], an approximate linearization approach has been used to analyze the convergence of the gradient-based algorithm under order-α IP criterion. Consider the ADALINE model:

$$\hat{y}_k = X_k^T W = \sum_{i=1}^{m} w_i x_{i,k} \tag{4.123}$$

where $X_k = [x_{1,k}, x_{2,k}, \ldots, x_{m,k}]^T$ is the m-dimensional input vector, $W = [w_1, w_2, \ldots, w_m]^T$ is the weight vector. The gradient based identification algorithm under order-α ($\alpha > 1$) information potential criterion can be expressed as

$$W_{k+1} = W_k + \eta \nabla \hat{V}_\alpha \tag{4.124}$$

where $\nabla \hat{V}_\alpha$ denotes the gradient of the empirical order-α IP with respect to the weight vector, which can be calculated as

$$\nabla \hat{V}_\alpha = \partial \hat{V}_\alpha / \partial W$$

$$= -\frac{(\alpha - 1)}{N^\alpha} \sum_{j=1}^{N} \left\{ \left(\sum_{i=1}^{N} K_h(e_j - e_i) \right)^{\alpha - 2} \sum_{i=1}^{N} (K'_h(e_j - e_i)(X_j - X_i)) \right\}$$

(4.125)

As the kernel function $K_h(.)$ satisfies $K_h(x) = K(x/h)/h$, we have

$$K'_h(x) = K'(x/h)/h^2$$

(4.126)

So, one can rewrite Eq. (4.125) as

$$\nabla \hat{V}_\alpha = -\frac{(\alpha - 1)}{h^\alpha N^\alpha} \sum_{j=1}^{N} \left\{ \left(\sum_{i=1}^{N} K(\Delta e_{ji}) \right)^{\alpha - 2} \sum_{i=1}^{N} (K'(\Delta e_{ji})(X_j - X_i)) \right\}$$ (4.127)

where $\Delta e_{ji} = (e_j - e_i)/h$.

If the weight vector W lies in the neighborhood of the optimal solution W^*, one can obtain a linear approximation of the gradient $\nabla \hat{V}_\alpha$ using the first-order Taylor expansion:

$$\nabla \hat{V}_\alpha(W) \approx \nabla \hat{V}_\alpha(W^*) + H \times (W - W^*) = H \times (W - W^*)$$ (4.128)

where $H = \partial \nabla \hat{V}_\alpha^T(W^*)/\partial W$ is the Hessian matrix, i.e.,

$$H = \partial \nabla \hat{V}_\alpha^T(W^*)/\partial W$$

$$= \frac{(\alpha - 1)}{h^\alpha N^\alpha} \sum_j \left[\sum_i K(\Delta e_{ji}(W^*)) \right]^{\alpha - 3}$$

$$\times \left\{ \begin{array}{l} (\alpha-2)\left[\sum_i K'(\Delta e_{ji}(W^*))(X_i - X_j)\right] \\ \times \left[\sum_i K'(\Delta e_{ji}(W^*))(X_i - X_j)^T\right] \\ + \left[\sum_i K(\Delta e_{ji}(W^*))\right] \\ \times \left[\sum_i K''(\Delta e_{ji}(W^*))(X_i - X_j)(X_i - X_j)^T\right] \end{array} \right\}$$

(4.129)

Substituting Eq. (4.128) into Eq. (4.124), and subtracting W^* from both sides, one obtains

$$\tilde{W}_{k+1} = \tilde{W}_k + \eta H \tilde{W}_k$$ (4.130)

where $\tilde{W}_k = W^* - W_k$ is the weight error vector. The convergence analysis of the above linear recursive algorithm is very simple. Actually, one can just borrow the well-known convergence analysis results from the LMS convergence theory [18−20]. Assume that the Hessian matrix H is a normal matrix and can be decomposed into the following normal form:

$$H = Q\Lambda Q^{-1} = Q\Lambda Q^T \tag{4.131}$$

where Q is an $m \times m$ orthogonal matrix, $\Lambda = \text{diag}[\lambda_1, \ldots, \lambda_m]$, λ_i is the eigenvalue of H. Then, the linear recursion (4.130) can be expressed as

$$\tilde{W}_{k+1} = Q(I + \eta\Lambda)Q^{-1}\tilde{W}_k \tag{4.132}$$

Clearly, if the following conditions are satisfied, the weight error vector $\tilde{W}_k$ will converge to the zero vector (or equivalently, the weight vector W_k will converge to the optimal solution):

$$|1 + \eta\lambda_i| < 1, \quad i = 1, \ldots, m, \tag{4.133}$$

Thus, a sufficient condition that ensures the convergence of the algorithm is

$$\begin{cases} \lambda_i < 0, i = 1, \ldots, m \\ 0 < \eta < 2/(\max_i |\lambda_i|) \end{cases} \tag{4.134}$$

Further, the approximate time constant corresponding to λ_i will be

$$\tau_i = \frac{-1}{\log(1 + \eta\lambda_i)} \approx \frac{1}{\eta|\lambda_i|} \tag{4.135}$$

In Ref. [102], it has also been proved that if the kernel width h increases, the absolute values of the eigenvalues will decrease, and the time constants will increase, that is, the convergence speed of the algorithm will become slower.

4.4.2 *Energy Conservation Relation*

Here, the energy conservation relation is not the well-known physical law of conservation of energy, but refers to a certain identical relation that exists among the WEP, *a priori* error, and *a posteriori* error in adaptive filtering algorithms (such as the LMS algorithm). This fundamental relation can be used to analyze the mean square convergence behavior of an adaptive filtering algorithm [196].

Given an adaptive filtering algorithm with error nonlinearity [29]:

$$W_{k+1} = W_k + \eta f(e_k)X_k \tag{4.136}$$

where $f(.)$ is the error nonlinearity function ($f(x) = x$ corresponds to the LMS algorithm), the following energy conservation relation holds:

$$E\left[\|\tilde{W}_{k+1}\|^2\right] + E\left[\frac{(e_k^a)^2}{\|X_k\|^2}\right] = E\left[\|\tilde{W}_k\|^2\right] + E\left[\frac{(e_k^p)^2}{\|X_k\|^2}\right] \tag{4.137}$$

where $E\left[\|\tilde{W}_k\|^2\right]$ is the WEP at instant k (or iteration k), e_k^a and e_k^p are, respectively, the *a priori* error and *a posteriori* error:

$$e_k^a \triangleq \tilde{W}_k^T X_k, \quad e_k^p \triangleq \tilde{W}_{k+1}^T X_k \tag{4.138}$$

One can show that the IG algorithms (assuming ADALINE model) also satisfy the energy conservation relation similar to Eq. (4.137) [106]. Let us consider the sliding information gradient algorithm (with ϕ entropy criterion):

$$W_{k+1} = W_k - \eta \frac{\partial}{\partial W} \hat{H}_{\phi,k}(e) \tag{4.139}$$

where $\hat{H}_{\phi,k}(e)$ is the empirical error entropy at iteration k:

$$\hat{H}_{\phi,k}(e) = \frac{1}{L} \sum_{j=1}^{L} \psi\left(\frac{1}{L} \sum_{i=1}^{L} K_h(e_{j,k} - e_{i,k})\right) \tag{4.140}$$

where $\psi(x) = \phi(x)/x$, $e_{i,k}$ is the i th error sample used to calculate the empirical error entropy at iteration k. Given the ϕ-function and the kernel function K_h, the empirical error entropy $\hat{H}_{\phi,k}(e)$ will be a function of the error vector $e_k = [e_{1,k}, e_{2,k}, \ldots, e_{L,k}]^T$, which can be written as

$$\hat{H}_{\phi,k}(e) = F_{\phi,K_h}(e_k) \tag{4.141}$$

This function will be continuously differentiable with respect to e_k, provided that both functions ϕ and K_h are continuously differentiable.

In ADALINE system identification, the error $e_{i,k}$ will be

$$e_{i,k} = z_{i,k} - \hat{y}_{i,k} = z_{i,k} - X_{i,k}W \tag{4.142}$$

where $z_{i,k}$ is the i th measurement at iteration k, $\hat{y}_{i,k} = X_{i,k}W$ is the i th model output at iteration k, $X_{i,k} = [x_{1,k}^{(i)}, x_{2,k}^{(i)}, \ldots, x_{m,k}^{(i)}]$ is the i th input vector[11] at iteration k, and

---

[11] Here the input vector is a row vector.

$W = [w_1, w_2, \ldots, w_m]^T$ is the ADALINE weight vector. One can write Eq. (4.142) in the form of block data, i.e.,

$$e_k = z_k - \hat{y}_k = z_k - \mathcal{X}_k W \tag{4.143}$$

where $z_k = [z_{1,k}, z_{2,k}, \ldots, z_{L,k}]^T$, $\hat{y}_k = [\hat{y}_{1,k}, \hat{y}_{2,k}, \ldots, \hat{y}_{L,k}]^T$, and $\mathcal{X}_k = [X_{1,k}^T, X_{2,k}^T, \ldots, X_{L,k}^T]^T$ ($L \times m$ input matrix).

The block data $\{\mathcal{X}_k, z_k, \hat{y}_k, e_k\}$ can be constructed in various manners. Two typical examples are

1. One time shift:

$$\begin{cases} \mathcal{X}_k = [X_k^T, X_{k-1}^T, \ldots, X_{k-L+1}^T]^T \\ z_k = [z_k, z_{k-1}, \ldots, z_{k-L+1}]^T \\ \hat{y}_k = [\hat{y}_k, \hat{y}_{k-1}, \ldots, \hat{y}_{k-L+1}]^T \\ e_k = [e_k, e_{k-1}, \ldots, e_{k-L+1}]^T \end{cases} \tag{4.144}$$

2. L-time shift:

$$\begin{cases} \mathcal{X}_k = [X_{kL}^T, X_{kL-1}^T, \ldots, X_{(k-1)L+1}^T]^T \\ z_k = [z_{kL}, z_{kL-1}, \ldots, z_{(k-1)L+1}]^T \\ \hat{y}_k = [\hat{y}_{kL}, \hat{y}_{kL-1}, \ldots, \hat{y}_{(k-1)L+1}]^T \\ e_k = [e_{kL}, e_{kL-1}, \ldots, e_{(k-1)L+1}]^T \end{cases} \tag{4.145}$$

Combining Eqs. (4.139), (4.141), and (4.143), we can derive (assuming ADALINE model)

$$\begin{aligned} W_{k+1} &= W_k - \eta \frac{\partial F_{\phi,K_h}(e_k)}{\partial W} \\ &= W_k - \eta \frac{\partial e_k^T}{\partial W} \frac{\partial F_{\phi,K_h}(e_k)}{\partial e_k} \\ &= W_k + \eta \mathcal{X}_k^T f(e_k) \end{aligned} \tag{4.146}$$

where $f(e_k) = [f_1(e_k), f_2(e_k), \ldots, f_L(e_k)]^T$, in which

$$f_i(e_k) = \frac{\partial F_{\phi,K_h}(e_k)}{\partial e_{i,k}} \tag{4.147}$$

With QIP criterion ($\phi(x) = -x^2$), the function $f_i(e_k)$ can be calculated as (assuming K_h is a Gaussian kernel):

$$
\begin{aligned}
f_i(e_k) &= \frac{\partial F_{\phi,K_h}(e_k)}{\partial e_{i,k}}\Big|_{\phi(x)=-x^2} = \frac{\partial\{-\hat{V}_2(e_k)\}}{\partial e_{i,k}} \\[2mm]
&= -\frac{1}{L^2}\frac{\partial}{\partial e_{i,k}}\left\{\sum_{i=1}^{L}\sum_{j=1}^{L}K_h(e_{i,k}-e_{j,k})\right\} \\[2mm]
&= -\frac{2}{L^2}\frac{\partial}{\partial e_{i,k}}\left\{\sum_{j=1}^{L}K_h(e_{i,k}-e_{j,k})\right\} \\[2mm]
&= \frac{2}{L^2 h^2}\left\{\sum_{j=1}^{L}(e_{i,k}-e_{j,k})K_h(e_{i,k}-e_{j,k})\right\}
\end{aligned}
\tag{4.148}
$$

The algorithm in Eq. (4.146) is in form very similar to the adaptive filtering algorithm with error nonlinearity, as expressed in Eq. (4.136). In fact, Eq. (4.146) can be regarded, to some extent, as a "block" version of Eq. (4.136). Thus, one can study the mean square convergence behavior of the algorithm (4.146) by similar approach as in mean square analysis of the algorithm (4.136). It should also be noted that the objective function behind algorithm (4.146) is not limited to the error entropy. Actually, the cost function $F_{\phi,K}(e_k)$ can be extended to any function of e_k, including the simple block mean square error (BMSE) criterion that is given by [197]

$$
\mathrm{BMSE} = \frac{1}{L}e_k^T e_k
\tag{4.149}
$$

We now derive the energy conservation relation for the algorithm (4.146). Assume that the unknown system and the adaptive model are both ADALINE structures with the same dimension of weights.[12] Let the measured output (in the form of block data) be

$$
z_k = \mathscr{X}_k W^* + v_k
\tag{4.150}
$$

where $W^* = [w_1^*, w_2^*, \ldots, w_m^*]^T$ is the weight vector of unknown system and $v_k = [v_{1,k}, v_{2,k}, \ldots, v_{L,k}]^T$ is the noise vector. In this case, the error vector e_k can be expressed as

$$
e_k = \mathscr{X}_k \tilde{W}_k + v_k
\tag{4.151}
$$

---

[12] This configuration has been widely adopted due to its convenience for convergence analysis (e.g., see Ref. [196]).

where $\tilde{W}_k = W^* - W_k$ is the weight error vector. In addition, we define the *a priori* and *a posteriori* error vectors e_k^a and e_k^p:

$$\begin{cases} e_k^a = [e_{1,k}^a, e_{2,k}^a, \ldots, e_{L,k}^a]^T = \mathscr{X}_k \tilde{W}_k \\ e_k^p = [e_{1,k}^p, e_{2,k}^p, \cdots, e_{L,k}^p]^T = \mathscr{X}_k \tilde{W}_{k+1} \end{cases} \tag{4.152}$$

Clearly, e_k^a and e_k^p have the following relationship:

$$e_k^p = e_k^a + \mathscr{X}_k(\tilde{W}_{k+1} - \tilde{W}_k) = e_k^a - \mathscr{X}_k(W_{k+1} - W_k) \tag{4.153}$$

Combining Eqs. (4.146) and (4.153) yields

$$e_k^p = e_k^a - \eta \mathscr{X}_k \mathscr{X}_k^T f(e_k) = e_k^a - \eta \mathscr{R}_k f(e_k) \tag{4.154}$$

where $\mathscr{R}_k = \mathscr{X}_k \mathscr{X}_k^T$ is an $L \times L$ symmetric matrix with elements $\mathscr{R}_k(ij) = X_{i,k} X_{j,k}^T$. Assume that the matrix $\mathscr{R}_k$ is invertible (i.e., $\det \mathscr{R}_k \neq 0$). Then we have

$$\begin{aligned} & e_k^p = e_k^a - \eta \mathscr{R}_k f(e_k) \\ & \Rightarrow \mathscr{R}_k^{-1}(e_k^p - e_k^a) = -\eta f(e_k) \\ & \Rightarrow \mathscr{X}_k^T \mathscr{R}_k^{-1}(e_k^p - e_k^a) = -\eta \mathscr{X}_k^T f(e_k) \\ & \Rightarrow \mathscr{X}_k^T \mathscr{R}_k^{-1}(e_k^p - e_k^a) = \tilde{W}_{k+1} - \tilde{W}_k \end{aligned} \tag{4.155}$$

And hence

$$\tilde{W}_{k+1} = \tilde{W}_k + \mathscr{X}_k^T \mathscr{R}_k^{-1}(e_k^p - e_k^a) \tag{4.156}$$

Both sides of Eq. (4.156) should have the same energy, i.e.,

$$\tilde{W}_{k+1}^T \tilde{W}_{k+1} = [\tilde{W}_k + \mathscr{X}_k^T \mathscr{R}_k^{-1}(e_k^p - e_k^a)]^T \times [\tilde{W}_k + \mathscr{X}_k^T \mathscr{R}_k^{-1}(e_k^p - e_k^a)] \tag{4.157}$$

From Eq. (4.157), after some simple manipulations, we obtain the following energy conservation relation:

$$\left\| \tilde{W}_{k+1} \right\|^2 + \left\| e_k^a \right\|_{\mathscr{R}_k^{-1}}^2 = \left\| \tilde{W}_k \right\|^2 + \left\| e_k^p \right\|_{\mathscr{R}_k^{-1}}^2 \tag{4.158}$$

where $\left\| \tilde{W}_k \right\|^2 = \tilde{W}_k^T \tilde{W}_k$, $\left\| e_k^a \right\|_{\mathscr{R}_k^{-1}}^2 = e_k^{aT} \mathscr{R}_k^{-1} e_k^a$, $\left\| e_k^p \right\|_{\mathscr{R}_k^{-1}}^2 = e_k^{pT} \mathscr{R}_k^{-1} e_k^p$. Further, in order to analyze the mean square convergence performance of the algorithm, we take the expectations of both sides of (4.158) and write

$$E\left[\left\| \tilde{W}_{k+1} \right\|^2 \right] + E\left[\left\| e_k^a \right\|_{\mathscr{R}_k^{-1}}^2 \right] = E\left[\left\| \tilde{W}_k \right\|^2 \right] + E\left[\left\| e_k^p \right\|_{\mathscr{R}_k^{-1}}^2 \right] \tag{4.159}$$

The energy conservation relation (4.159) shows how the energies (powers) of the error quantities evolve in time, which is exact for any adaptive algorithm described in Eq. (4.146), and is derived without any approximation and assumption (except for the condition that $\mathscr{R}_k$ is invertible). One can also observe that Eq. (4.159) is a generalization of Eq. (4.137) in the sense that the *a priori* and *a posteriori* error quantities are extended to vector case.

4.4.3 Mean Square Convergence Analysis Based on Energy Conservation Relation

The energy conservation relation (4.159) characterizes the evolution behavior of the weight error power (WEP). Substituting $e_k^p = e_k^a - \eta \mathscr{R}_k f(e_k)$ into (4.159), we obtain

$$E\left[\left\|\tilde{W}_{k+1}\right\|^2\right] = E\left[\left\|\tilde{W}_k\right\|^2\right] - 2\eta E[e_k^{aT} f(e_k)] + \eta^2 E[f^T(e_k)\mathscr{R}_k f(e_k)] \qquad (4.160)$$

To evaluate the expectations $E[e_k^{aT} f(e_k)]$ and $E[f^T(e_k)\mathscr{R}_k f(e_k)]$, some assumptions are given below [106]:

- **A1**: The noise $\{v_k\}$ is independent, identically distributed, and independent of the input $\{X_k\}$;
- **A2**: The *a priori* error vector e_k^a is jointly Gaussian distributed;
- **A3**: The input vectors $\{X_k\}$ are zero-mean independent, identically distributed;
- **A4**: $\forall\, i,j \in \{1,\ldots,L\}$, $\mathscr{R}_k(ij)$ and $\{e_{i,k}, e_{j,k}\}$ are independent.

Remark: Assumptions A1 and A2 are popular and have been widely used in convergence analysis for many adaptive filtering algorithms [196]. As pointed out in [29], the assumption A2 is reasonable for longer weight vector by central limit theorem arguments. The assumption A3 restricts the input sequence $\{X_k\}$ to white regression data, which is also a common practice in the literature (e.g., as in Refs. [28,198,199]). The assumption A4 is somewhat similar to the uncorrelation assumption in Ref. [29], but it is a little stronger. This assumption is reasonable under assumptions A1 and A3, and will become more realistic as the weight vector gets longer (justified by the law of large numbers).

In the following, for tractability, we only consider the case in which the block data are constructed by "*L*-time shift" approach (see Eq. (4.145)). In this case, the assumption A3 implies: (i) the input matrices $\{\mathscr{X}_k\}$ are zero mean, independent, identically distributed and (ii) $\mathscr{X}_k$ and $\tilde{W}_k$ are mutually independent. Combining assumption A3 and the independence between $\mathscr{X}_k$ and $\tilde{W}_k$, one may easily conclude that the components of the *a priori* error vector e_k^a are also zero-mean, independent, identically distributed. Thus, by Gaussian assumption A2, the PDF of e_k^a can be expressed as

$$p_{e_k^a}(e_k^a) = \left(\frac{1}{\sqrt{2\pi\gamma_k^2}}\right)^L \prod_{i=1}^{L} \exp\left(-\frac{(e_{i,k}^a)^2}{2\gamma_k^2}\right) \qquad (4.161)$$

where $\gamma_k^2 = E[(e_{i,k}^a)^2]$ is the *a priori* error power. Further, by assumption A1, the *a priori* error vector e_k^a and the noise vector v_k are independent, and hence

$$
\begin{aligned}
E[e_k^{aT} f(e_k)] &= E[e_k^{aT} f(e_k^a + v_k)] \\
&= \int e_k^{aT} f(e_k^a + v_k) p_{v_k}(v_k) dv_k p_{e_k^a}(e_k^a) de_k^a \\
&= \left(\frac{1}{\sqrt{2\pi\gamma_k^2}}\right)^L \int p_{v_k}(v_k) dv_k \int e_k^{aT} f(e_k^a + v_k) \times \prod_{i=1}^{L} \exp\left(-\frac{(e_{i,k}^a)^2}{2\gamma_k^2}\right) de_k^a
\end{aligned}
$$

$$(4.162)$$

where $p_{v_k}(.)$ denotes the PDF of v_k. The inner integral depends on e_k^a through the second moment γ_k^2 only, and so does $E\{e_k^{aT} f(e_k)\}$. Thus, given the noise distribution $p_{v_k}(.)$, the expectation $E\{e_k^{aT} f(e_k)\}$ can be expressed as a function of γ_k^2, which enables us to define the following function[13] :

$$
h_G(\gamma_k^2) \triangleq E[e_k^{aT} f(e_k)]/\gamma_k^2 \tag{4.163}
$$

It follows that

$$
E[e_k^{aT} f(e_k)] = \gamma_k^2 h_G(\gamma_k^2) \tag{4.164}
$$

Next, we evaluate the expectation $E[f^T(e_k)\mathcal{R}_k f(e_k)]$. As $\mathcal{R}_k(ij) = X_{i,k} X_{j,k}^T$, $e_k = \mathcal{X}_k \tilde{W}_k + v_k$, by assumptions A1, A3, and A4, $\mathcal{R}_k(ij)$ and e_k will be independent. Thus

$$
\begin{aligned}
E[f^T(e_k)\mathcal{R}_k f(e_k)] &= \sum_{i=1}^{L} \sum_{j=1}^{L} E[f_i(e_k) f_j(e_k) \mathcal{R}_k(ij)] \\
&= \sum_{i=1}^{L} \sum_{j=1}^{L} E[f_i(e_k) f_j(e_k)] E[\mathcal{R}_k(ij)] \\
&\overset{(a)}{=} \sum_{i=1}^{L} E[f_i^2(e_k)] E[\mathcal{R}_k(ii)] \\
&\overset{(b)}{=} \sum_{i=1}^{L} E[f_i^2(e_k)] E\left[\|X_k\|^2\right]
\end{aligned}
$$

$$(4.165)$$

[13] Similar to [29], the subscript G in h_G indicates that the Gaussian assumption A2 is the main assumption leading to the defined expression (4.163). The subscript I for h_I, which is defined later in (4.166), however, suggests that the independence assumption A4 is the major assumption in evaluating the expectation.

where (a) and (b) follow from the assumption A3. Since e_k^a is Gaussian and independent of the noise vector v_k, the term $E\{f_i^2(e_k)\}$ will also depend on e_k^a through γ_k^2 only, and this prompts us to define the function h_I:

$$h_I(\gamma_k^2) \triangleq \sum_{i=1}^{L} E[f_i^2(e_k)] \tag{4.166}$$

Combining Eqs. (4.165) and (4.166) yields

$$E[f^T(e_k)\mathscr{R}_k f(e_k)] = h_I(\gamma_k^2)E\left[\left\|X_k\right\|^2\right] \tag{4.167}$$

Substituting Eqs. (4.164) and (4.167) into Eq. (4.160), we obtain

$$E\left[\left\|\tilde{W}_{k+1}\right\|^2\right] = E\left[\left\|\tilde{W}_k\right\|^2\right] - 2\eta\gamma_k^2 h_G(\gamma_k^2) + \eta^2 h_I(\gamma_k^2)E\left[\left\|X_k\right\|^2\right] \tag{4.168}$$

Now, we use the recursion formula (4.168) to analyze the mean square convergence behavior of the algorithm (4.146), following the similar derivations in Ref. [29].

4.4.3.1 Sufficient Condition for Mean Square Convergence

From Eq. (4.168), it is easy to observe

$$\begin{aligned} &E\left[\left\|\tilde{W}_{k+1}\right\|^2\right] \leq E\left[\left\|\tilde{W}_k\right\|^2\right] \\ &\Leftrightarrow -2\eta\gamma_k^2 h_G(\gamma_k^2) + \eta^2 h_I(\gamma_k^2)E\left[\left\|X_k\right\|^2\right] \leq 0 \end{aligned} \tag{4.169}$$

Therefore, if we choose the step-size η such that for all k

$$\eta \leq \frac{2\gamma_k^2 h_G(\gamma_k^2)}{h_I(\gamma_k^2)E\left[\left\|X_k\right\|^2\right]} \tag{4.170}$$

then the sequence of WEP $\left\{E\left[\left\|\tilde{W}_k\right\|^2\right]\right\}$ will be monotonically decreasing (and hence convergent). Thus, a sufficient condition for the mean square convergence of the algorithm (4.146) would be

$$\begin{aligned} \eta &\leq \inf_{\gamma_k^2 \in \Omega} \frac{2\gamma_k^2 h_G(\gamma_k^2)}{h_I(\gamma_k^2)E\left[\left\|X_k\right\|^2\right]} \\ &\stackrel{(a)}{=} \frac{2}{E\left[\left\|X_k\right\|^2\right]} \inf_{\gamma_k^2 \in \Omega} \left\{\frac{\gamma_k^2 h_G(\gamma_k^2)}{h_I(\gamma_k^2)}\right\} \\ &= \frac{2}{E\left[\left\|X_k\right\|^2\right]} \inf_{\gamma_k^2 \in \Omega} \rho(\gamma_k^2) \end{aligned} \tag{4.171}$$

where (a) comes from the assumption that the input vector is stationary, $\rho(x) = xh_G(x)/h_I(x)$, Ω denotes the set of all possible values of γ_k^2 ($k \geq 0$).

In general, the above upper bound of the step-size is conservative. However, if $\rho(x)$ is a monotonically increasing function over $[0, \infty)$, i.e., $\forall x_1 > x_2 \geq 0$, $\rho(x_1) \geq \rho(x_2)$, one can explicitly derive the maximum value (tight upper bound) of the step-size that ensures the mean square convergence. Let's come back to the condition in Eq. (4.170) and write

$$\eta \leq \frac{2\rho(\gamma_k^2)}{E[\|X_k\|^2]}, \quad \forall k \geq 0 \tag{4.172}$$

Under this condition, the WEP will be monotonically decreasing. As the *a priori* error power γ_k^2 is proportional to the WEP $E\left[\|\tilde{W}_k\|^2\right]$ (see Eq. (4.177)), in this case, the *a priori* error power will also be monotonically decreasing, i.e.,

$$\gamma_0^2 \geq \gamma_1^2 \geq \cdots \geq \gamma_k^2 \geq \gamma_{k+1}^2 \geq \cdots \tag{4.173}$$

where γ_0^2 is the initial *a priori* error power. So, the maximum step-size is

$$\eta_{\max} = \max\left\{\eta:0 < \eta \leq \frac{2\rho(\gamma_k^2)}{E[\|X_k\|^2]}, \quad \forall k \geq 0\right\}$$

$$= \max\left\{\eta:0 < \eta \leq \frac{2\rho(\gamma_k^2)}{E[\|X_k\|^2]}, \gamma_k^2 \leq \gamma_0^2\right\} \tag{4.174}$$

$$\overset{(a)}{=} \frac{2\rho(\gamma_0^2)}{E[\|X_k\|^2]}$$

where (a) follows from Eq. (4.173) and the monotonic property of $\rho(x)$. This maximum step-size depends upon the initial *a priori* error power. When $\eta = \eta_{\max}$, we have

$$\begin{cases} E[\|\tilde{W}_k\|^2] = E[\|\tilde{W}_0\|^2], & \forall k \geq 0 \\ \gamma_k^2 = \gamma_0^2 \end{cases} \tag{4.175}$$

In this case, the learning is at the edge of convergence (WEP remains constant).

Remark If the step-size η is below the upper bound or smaller than the maximum value $\eta_{\max}$, the WEP will be decreasing. However, this does not imply that the WEP will converge to zero. There are two reasons for this. First, for a stochastic gradient-based algorithm, there always exist some excess errors (misadjustments). Second, the algorithm may converge to a local minimum (if any).

4.4.3.2 Mean Square Convergence Curve

One can establish a more homogeneous form for the recursion formula (4.168). First, it is easy to derive

$$
\begin{aligned}
\gamma_k^2 &= E[(e_{i,k}^a)^2] = E[(X_{i,k}\tilde{W}_k)^2] \\
&= E[\tilde{W}_k^T(X_{i,k}^T X_{i,k})\tilde{W}_k] \\
&\overset{(a)}{=} E[\tilde{W}_k^T(E[X_{i,k}^T X_{i,k}])\tilde{W}_k] \\
&= E[\|\tilde{W}_k\|_{R_X}^2]
\end{aligned}
\tag{4.176}
$$

where (a) follows from the independence between $\mathscr{X}_k$ and $\tilde{W}_k$, $\|\tilde{W}_k\|_{R_X}^2 = \tilde{W}_k^T R_X \tilde{W}_k$, $R_X = E[X_{i,k}^T X_{i,k}]$. As the input data are assumed to be zero-mean, independent, identically distributed, we have $R_X = \sigma_x^2 I$ (I is an $m \times m$-dimensional unit matrix), and hence

$$
\gamma_k^2 = \sigma_x^2 E[\|\tilde{W}_k\|^2]
\tag{4.177}
$$

Substituting Eq. (4.177) and $E[\|X_k\|^2] = m\sigma_x^2$ into Eq. (4.168) yields the equation that governs the mean square convergence curve:

$$
E[\|\tilde{W}_{k+1}\|^2] = E[\|\tilde{W}_k\|^2] - 2\eta\sigma_x^2 E[\|\tilde{W}_k\|^2]h_G(\sigma_x^2 E[\|\tilde{W}_k\|^2]) + m\eta^2\sigma_x^2 h_I(\sigma_x^2 E[\|\tilde{W}_k\|^2])
\tag{4.178}
$$

4.4.3.3 Mean Square Steady-State Performance

We can use Eq. (4.178) to evaluate the mean square steady-state performance. Suppose the WEP reaches a steady-state value, i.e.,

$$
\lim_{k\to\infty} E[\|\tilde{W}_{k+1}\|^2] = \lim_{k\to\infty} E[\|\tilde{W}_k\|^2]
\tag{4.179}
$$

Then the mean square convergence equation (4.178) becomes, in the limit

$$
\lim_{k\to\infty} E[\|\tilde{W}_k\|^2]h_G(\sigma_x^2 E[\|\tilde{W}_k\|^2]) = \lim_{k\to\infty} \frac{m\eta}{2} h_I(\sigma_x^2 E[\|\tilde{W}_k\|^2])
\tag{4.180}
$$

It follows that

$$
\lim_{k\to\infty} E[\|\tilde{W}_k\|^2]h_G(\sigma_x^2 \lim_{k\to\infty} E[\|\tilde{W}_k\|^2]) = \frac{m\eta}{2} h_I(\sigma_x^2 \lim_{k\to\infty} E[\|\tilde{W}_k\|^2])
\tag{4.181}
$$

Denote S_{WEP} the steady-state WEP, i.e., $S_{\text{WEP}} = \lim\limits_{k \to \infty} E[\|\tilde{W}_k\|^2]$, we have

$$S_{\text{WEP}} = \frac{m\eta h_I(\sigma_x^2 S_{\text{WEP}})}{2h_G(\sigma_x^2 S_{\text{WEP}})} \tag{4.182}$$

Therefore, if the adaptive algorithm (4.146) converges, the steady-state WEP S_{WEP} will be a positive solution of Eq. (4.182), or equivalently, S_{WEP} will be a positive FP of the function $\varphi(\xi) = m\eta h_I(\sigma_x^2 \xi)/\{2h_G(\sigma_x^2 \xi)\}$.

Further, denote S_{EMSE} the steady-state excess mean square error (EMSE), i.e., $S_{\text{EMSE}} = \lim_{k \to \infty} \gamma_k^2$. By Eq. (4.177), we can easily evaluate S_{EMSE} as

$$S_{\text{EMSE}} = \lim_{k \to \infty} \sigma_x^2 E\left[\|\tilde{W}_k\|^2\right] = \sigma_x^2 S_{\text{WEP}} \tag{4.183}$$

The steady-state EMSE is in linear proportion to the steady-state WEP.

So far we have derived the mean square convergence performance for adaptive algorithm (4.146), under the assumptions A1−A4. The derived results depend mainly on two functions: $h_G(.)$, $h_I(.)$. In the following, we will derive the exact expressions for the two functions for QIP criterion ($\phi(x) = -x^2$).

By Eq. (4.148), under QIP criterion we have

$$f_i(e_k) = \frac{2}{L^2 h^2}\left\{\sum_{j=1}^{L}(e_{i,k} - e_{j,k})K_h(e_{i,k} - e_{j,k})\right\} \tag{4.184}$$

Hence, the function $h_G(\gamma_k^2)$ can be expressed as

$$\begin{aligned}
h_G(\gamma_k^2) &= E[e_k^{aT}f(e_k)]/\gamma_k^2 \\
&= \frac{1}{\gamma_k^2}\sum_{i=1}^{L}E[e_{i,k}^a f_i(e_k)] \\
&= \frac{2}{\gamma_k^2 L^2 h^2}\sum_{i=1}^{L}E\left[e_{i,k}^a \sum_{j=1}^{L}(e_{i,k} - e_{j,k})K_h(e_{i,k} - e_{j,k})\right] \\
&\overset{(a)}{=} \frac{2}{\gamma_k^2 L^2 h^2}\sum_{i=1}^{L}\{(L-1)E[e_{1,k}^a(e_{1,k} - e_{2,k})K_h(e_{1,k} - e_{2,k})]\} \\
&= \frac{2(L-1)}{\gamma_k^2 L h^2}E\left[e_{1,k}^a(e_{1,k} - e_{2,k})K_h(e_{1,k} - e_{2,k})\right]
\end{aligned} \tag{4.185}$$

where (a) comes from the fact that the error pairs $\{(e_{j,k}^a, e_{j,k}), j = 1, \ldots, L\}$ are independent, identically distributed. In addition, substituting (4.184) into (4.166) yields

$$h_I(\gamma_k^2) = \frac{4}{L^4 h^4} \sum_{i=1}^{L} E\left[\left(\sum_{j=1}^{L}(e_{i,k}-e_{j,k})K_h(e_{i,k}-e_{j,k})\right)^2\right]$$

$$\stackrel{(b)}{=} \frac{4}{L^4 h^4} \sum_{i=1}^{L} E\left[\left(\sum_{j=1}^{L}(e_{1,k}-e_{j,k})K_h(e_{1,k}-e_{j,k})\right)^2\right] \tag{4.186}$$

$$\stackrel{(c)}{=} \frac{4}{L^3 h^4}\left\{\begin{array}{l}(L-1)E[(e_{1,k}-e_{2,k})^2 K_h^2(e_{1,k}-e_{2,k})]+(L-1)(L-2)E \\ [(e_{1,k}-e_{2,k})(e_{1,k}-e_{3,k})K_h(e_{1,k}-e_{2,k})K_h(e_{1,k}-e_{3,k})]\end{array}\right\}$$

where (b) and (c) follow from the fact that the error samples $\{e_{j,k}, j=1,\ldots,L\}$ are independent, identically distributed.

In (4.185) and (4.186), the functions $h_G(\gamma_k^2)$ and $h_I(\gamma_k^2)$ do not yet have the explicit expressions in terms of the argument γ_k^2. In order to obtain the explicit expressions, one has to calculate the involved expectations using the PDFs of the *a priori* error and the noise. The calculation is rather complex and tedious. In the following, we only present the results for the Gaussian noise case.

Let the noise $\{v_k\}$ be a zero-mean white Gaussian process with variance λ^2. Then the error e_k will also be zero-mean Gaussian distributed with variance $\zeta_k^2 = \gamma_k^2 + \lambda^2$. In this case, one can derive

$$\begin{cases}h_G(\gamma_k^2) = \dfrac{2(L-1)(\gamma_k^4+(\lambda^2+2h^2)\gamma_k^2+h^2\lambda^2+h^4)}{L\sqrt{2\pi}(\gamma_k^2+h^2)(\zeta_k^2+h^2)(2\zeta_k^2+h^2)^{3/2}} \\[4mm] h_I(\gamma_k^2) = \dfrac{4(L-1)\zeta_k^2}{L^3\pi h^3(4\zeta_k^2+h^2)^{3/2}} + \\[4mm] \qquad \dfrac{2(L-1)(L-2)\zeta_k^2(4\zeta_k^8+16h^2\zeta_k^6+17h^4\zeta_k^4+7h^6\zeta_k^2+h^8)}{L^3\pi(2\zeta_k^2+h^2)^2(\zeta_k^4+3h^2\zeta_k^2+h^4)(3\zeta_k^4+4h^2\zeta_k^2+h^4)^{3/2}}\end{cases} \tag{4.187}$$

Substituting the explicit expressions in Eq. (4.187) into Eq. (4.178), one may obtain the exact convergence curve of the WEP, which can be described by a nonlinear dynamic system:

$$E\left[\|\tilde{W}_{k+1}\|^2\right] = h\left(E\left[\|\tilde{W}_k\|^2\right]\right) \tag{4.188}$$

where the function $h(\xi) = \xi - 2\eta\sigma_x^2\xi h_G(\sigma_x^2\xi) + m\eta^2\sigma_x^2 h_I(\sigma_x^2\xi)$.

In the following, a Monte Carlo simulation example is presented to verify the previous theoretical analysis results [106]. Consider the case in which the input signal and additive noise are both white Gaussian processes with unit power. Assume the unknown and adaptive systems are both ADALINE structures with weight

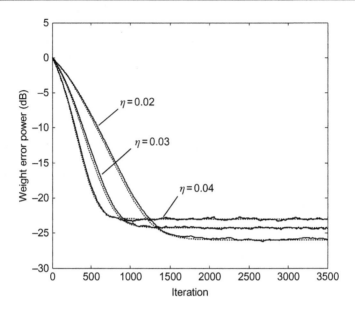

Figure 4.14 Simulated and theoretical learning curves for different step-sizes ($h = 1.0$, $L = 20$).

vector of length 25. The initial weight vector of the adaptive system was obtained by perturbing each coefficient of the ideal weight vector W^* by a random variable that is zero-mean Gaussian distributed with variance 0.04 (hence the initial WEP is 1.0 or 0 dB).

First, we examine the mean square convergence curves of the adaptive algorithm. For different values of the step-size η, kernel width h, and sliding data length L, the average convergence curves (solid) over 100 Monte Carlo runs and the corresponding theoretical learning curves (dotted) are plotted in Figures 4.14–4.16. Clearly, the experimental and theoretical results agree very well. Second, we verify the steady-state performance. As shown in Figures 4.17–4.19, the steady-state EMSEs generated by simulations match well with those calculated by theory. These simulated and theoretical results also demonstrate how the step-size η, kernel width h, and sliding data length L affect the performance of the adaptation: (i) a larger step-size produces a faster initial convergence, but results in a larger misadjustment; (ii) a larger kernel width causes a slower initial convergence, but yields a smaller misadjustment; (iii) a larger sliding data length achieves a faster initial convergence and a smaller misadjustment.[14]

In addition, we verify the upper bound on step-sizes that guarantee the convergence of the learning. For the case in which the kernel width $h = 0.2$, and the sliding data length $L = 20$, we plot in Figure 4.20 the curve of the function

---

[14] Increasing the sliding data length can improve both convergence speed and steady-state performance, however, this will increase dramatically the computational burden ($O(L^2)$).

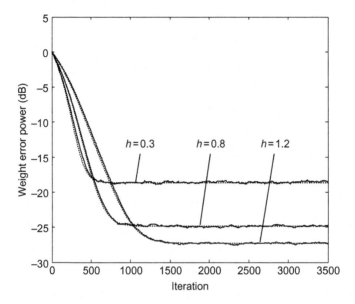

Figure 4.15 Simulated and theoretical learning curves for different kernel widths ($\eta = 0.03$, $L = 30$).

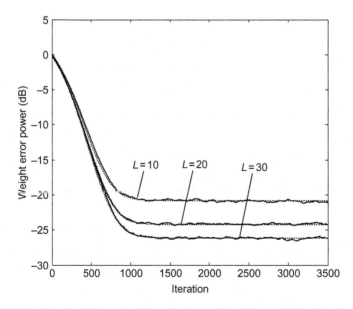

Figure 4.16 Simulated and theoretical learning curves for different sliding data lengths ($\eta = 0.03$, $h = 1.0$).

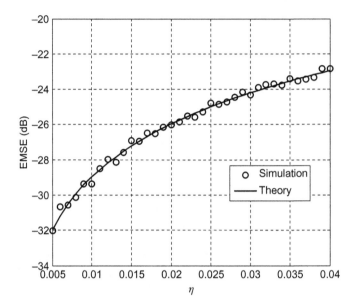

Figure 4.17 Simulated and theoretical EMSE versus step-size η ($h = 1.0$, $L = 20$).

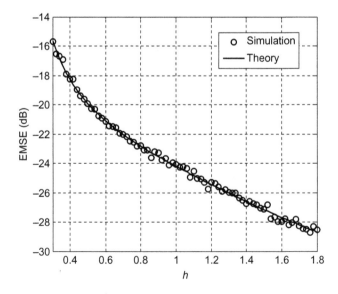

Figure 4.18 Simulated and theoretical EMSE versus kernel width h ($h = 0.03$, $L = 20$).

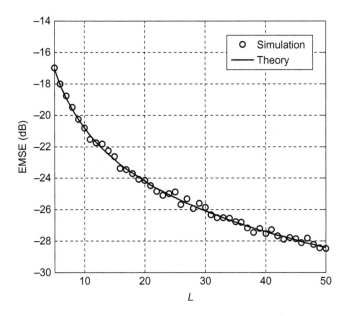

Figure 4.19 Simulated and theoretical EMSE versus sliding data length L ($\eta = 0.03$, $h = 1.0$).

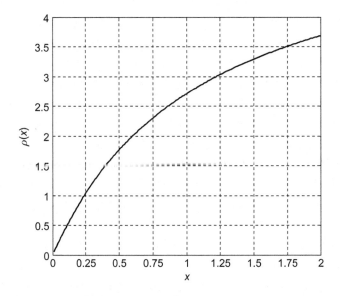

Figure 4.20 The function $\rho(x)$ ($h = 0.2$, $L = 20$).

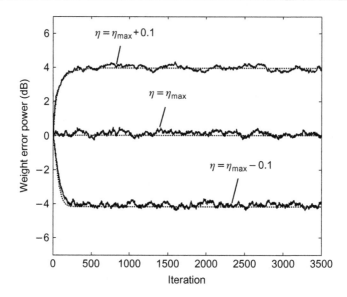

Figure 4.21 Simulated (solid) and theoretical (dotted) learning curves for step-sizes around η_{max}.

$\rho(x) = x h_G(x)/h_I(x)$. Clearly, this function is monotonically increasing. Thus by (4.174), we can calculate the maximum step-size $\eta_{\text{max}} \approx 0.217$. For step-sizes around $\eta_{\text{max}}(\eta_{\text{max}}, \eta_{\text{max}} \pm 0.1)$ the simulated and theoretical learning curves are shown in Figure 4.21. As expected, when $\eta = \eta_{\text{max}}$, the learning is at the edge of convergence. If η is above (or below) the maximum step-size η_{max}, the weight error power will be increasing (or decreasing).

4.5   Optimization of ϕ-Entropy Criterion

The ϕ-entropy criterion is very flexible. In fact, many entropy definitions are special cases of the ϕ-entropy. This flexibility, however, also brings the problem of how to select a good ϕ function to maximize the performance of the adaptation algorithm [200].

The selection of the ϕ function is actually an optimization problem. Denote J a quantitative performance index (convergence speed, steady-state accuracy, etc.) of the algorithm that we would like to optimize. Then the optimal ϕ function will be

$$\phi_{\text{opt}} = \arg \max_{\phi \in \Phi} J \tag{4.189}$$

where Φ represents the set of all possible ϕ functions. Due to the fact that different identification scenarios usually adopt different performance indexes, the above optimization problem must be further formulated in response to a specific identification system. In the following, we will focus on the FIR system identification, and assume for simplicity that the unknown system and the adaptive model are both FIR filters of the same order.

Before proceeding, we briefly review the optimization of the general error criterion $E[l(e)]$. In the adaptive filtering literature, there are mainly two approaches for the optimization of the general (usually non-MSE) error criterion. One approach regards the choice of the error criterion (or the l function) as a parameter search, in which a suitable structure of the criterion is assumed [26,201]. Such a design method usually leads to a suboptimal algorithm since the criterion is limited to a restricted class of functions. Another approach is proposed by Douglas and Meng [28], where the calculus of variations is used, and no prior information about the structure of error criterion is assumed. According to Douglas' method, in FIR system identification one can use the following performance index to optimize the error criterion:

$$J = - trE[\tilde{W}_{k+1}\tilde{W}_{k+1}^T] \tag{4.190}$$

where $\tilde{W}_{k+1}$ is the weight error vector at $k+1$ iteration. With the above performance index, the optimization of the l function can be formulated as the following optimization problem [28]:

$$\begin{cases} \min_\xi \int_{-\infty}^{+\infty} \xi^2(e)p_k(e)\mathrm{d}e \\ \text{s.t.} \int_{-\infty}^{+\infty} 2(\xi'(e) - \eta\lambda\{(\xi'(e))^2 + \xi(e)\xi''(e)\})p_k(e)\mathrm{d}e = 1 \end{cases} \tag{4.191}$$

where $\xi(e) = l'(e)$, $p_k(e)$ is the error PDF at k iteration, η is the step-size, and λ is the input signal power. By calculus of variations, one can obtain the optimal ξ function [28]:

$$\xi_{\mathrm{opt}}(e) = - \frac{p'_k(e)}{p_k(e) + \eta\lambda p''_k(e)} \tag{4.192}$$

The optimal l function can thus be expressed in the form of indefinite integral:

$$l_{\mathrm{opt}}(e) = \int \xi_{\mathrm{opt}}(e)\mathrm{d}e = \int - \frac{p'_k(e)}{p_k(e) + \eta\lambda p''_k(e)}\mathrm{d}e \tag{4.193}$$

which depends crucially on the error PDF $p_k(e)$.

Next, we will utilize Eq. (4.193) to derive an optimal ϕ-entropy criterion [200]. Assume that the error PDF $p_k(.)$ is symmetric, continuously differentiable (up to the second order), and unimodal with a peak at the origin. Then $p_k(.)$ satisfies: (i) invertible over interval $[0, +\infty)$ and (ii) $p''_k(.)$ is symmetric. Therefore, we have

$$p''_k(e) = p''_k(|e|) = p''_k(p_k^{-1}[p_k(e)]) = \beta(p_k(e)) \tag{4.194}$$

where $p_k^{-1}(.)$ denotes the inverse function of $p_k(.)$ over $[0, +\infty)$ and $\beta = p_k'' \circ p_k^{-1}$. So, we can rewrite Eq. (4.193) as

$$l_{opt}(e) = \int -\frac{p'_k(e)}{p_k(e) + \eta\lambda\beta(p_k(e))} de \tag{4.195}$$

Let the optimal ϕ-entropy criterion equal to the optimal error criterion $E[l_{opt}(e)]$, we have

$$
\begin{aligned}
H_{\phi_{opt}}(e) &= \int_{-\infty}^{\infty} \phi_{opt}[p_k(e)]de = E[l_{opt}(e)] \\
&= \int_{-\infty}^{\infty} l_{opt}(e)p_k(e)de \\
&= \int_{-\infty}^{\infty} \left\{ \int -\frac{p'_k(e)}{p_k(e) + \eta\lambda\beta(p_k(e))} de \right\} p_k(e)de \\
&= \int_{-\infty}^{\infty} \left\{ \int -\frac{1}{p_k(e) + \eta\lambda\beta(p_k(e))} dp_k(e) \right\} p_k(e)de
\end{aligned}
\tag{4.196}
$$

Hence

$$\phi_{opt}[p_k(e)] = \left\{ \int -\frac{1}{p_k(e) + \eta\lambda\beta(p_k(e))} dp_k(e) \right\} p_k(e) \tag{4.197}$$

Let $p_k(e) = x$, we obtain

$$\phi_{opt}(x) = \left\{ \int -\frac{1}{x + \eta\lambda\beta(x)} dx \right\} x \tag{4.198}$$

To achieve an explicit form of the function $\phi_{opt}(x)$, we consider a special case in which the error is zero-mean Gaussian distributed:

$$p_k(e) = \frac{1}{\sqrt{2\pi}\sigma_k} \exp\left(-\frac{e^2}{2\sigma_k^2}\right) \tag{4.199}$$

Then we have

$$p_k''(e) = \beta[p_k(e)] = \frac{\left\{\left(-1 - 2\log\sqrt{2\pi\sigma_k^2}\right)p_k(e)\right\} - \{2p_k(e)\log p_k(e)\}}{\sigma_k^2} \tag{4.200}$$

It follows that

$$\beta(x) = \frac{\left\{\left(-1 - 2\log\sqrt{2\pi\sigma_k^2}\right)x\right\} - \{2x\log x\}}{\sigma_k^2} \tag{4.201}$$

Substituting Eq. (4.201) into Eq. (4.198) yields

$$\phi_{opt}(x) = \left\{ \int \frac{-1}{\gamma_1 x + \gamma_2 x \log x} dx \right\} x = -\frac{x}{\gamma_2} \log(\gamma_1 + \gamma_2 \log x) + cx \qquad (4.202)$$

where $\gamma_1 = 1 - \eta\lambda\left(\left(1 + 2\log\sqrt{2\pi\sigma_k^2}\right)/\sigma_k^2\right)$, $\gamma_2 = -(2\eta\lambda/\sigma_k^2)$, and $c \in \mathbb{R}$ is a constant. Thus, we obtain the following optimal ϕ-entropy:

$$H_{\phi_{opt}}(e) = \int_{-\infty}^{+\infty} \left(-\frac{1}{\gamma_2} p(e)\log(\gamma_1 + \gamma_2\log p(e)) + cp(e) \right) de \qquad (4.203)$$

When $c = \left(1 + 2\log\sqrt{2\pi\sigma_k^2}\right)/2$, and $\eta \to 0$, we have

$$\lim_{\eta \to 0} \phi_{opt}(x) = \lim_{\eta \to 0} \left(-\frac{x}{\gamma_2}\log(\gamma_1 + \gamma_2\log x) + cx \right)$$

$$= \lim_{\eta \to 0} \frac{x\sigma_k^2}{2\eta\lambda}\log\left(1 - \eta\lambda\frac{1 + 2\log\sqrt{2\pi\sigma_k^2}}{\sigma_k^2} - \frac{2\eta\lambda}{\sigma_k^2}\log x\right) + cx$$

$$= \lim_{\eta \to 0} \frac{-x\sigma_k^2}{2\lambda} \cdot \frac{\lambda\frac{1 + 2\log\sqrt{2\pi\sigma_k^2}}{\sigma_k^2} + \frac{2\lambda}{\sigma_k^2}\log x}{1 - \eta\lambda\frac{1 + 2\log\sqrt{2\pi\sigma_k^2}}{\sigma_k^2} - \frac{2\eta\lambda}{\sigma_k^2}\log x} + cx$$

$$= \frac{-x}{2}\left(1 + 2\log\sqrt{2\pi\sigma_k^2} + 2\log x\right) + cx$$

$$= -x\log x$$

$$(4.204)$$

That is, as $\eta \to 0$ the derived optimal entropy will approach the Shannon entropy. One may therefore conclude that, under slow adaptation condition (η is small enough), Shannon's entropy is actually a suboptimal entropy criterion. Figure 4.22 shows the optimal ϕ functions for different step-sizes η (assume $\lambda = \sigma_k^2 = 1$).

One can easily derive the IG algorithms under the optimal ϕ-entropy criterion. For example, substituting Eq. (4.202) into Eq. (4.79), we have the following SIG algorithm:

$$W_k = W_{k-1} - \eta \frac{\sum_{i=k-L+1}^{k} \left\{ K'_h(e_k - e_i)\left(\frac{\partial \hat{y}_k}{\partial W} - \frac{\partial \hat{y}_i}{\partial W}\right) \right\}}{L(\gamma_1\hat{p}(e_k) + \gamma_2\hat{p}(e_k)\log\hat{p}(e_k))} \qquad (4.205)$$

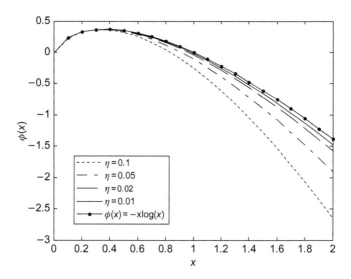

Figure 4.22 Optimal ϕ functions for different step-sizes η (adopted from Ref. [200]).

It is worth noting that the above optimal ϕ entropy was derived as an example for a restricted situation, in which the constraints include FIR identification, white Gaussian input and noise, etc. For more general situations, such as nonlinear and non-Gaussian cases, the derived ϕ function would no longer be an optimal one.

As pointed out in Ref. [28], the implementation of the optimal nonlinear error adaptation algorithm requires the exact knowledge of the noise's or error's PDFs. This is usually not the case in practice, since the characteristics of the noise or error may only be partially known or time-varying. In the implementation of the algorithm under the optimal ϕ-entropy criterion, however, the required PDFs are estimated by a nonparametric approach (say the KDE), and hence we don't need such *a priori* information. It must be noted that in Eq. (4.205), the parameters γ_1 and γ_2 are both related to the error variance σ_k^2, which is always time-varying during the adaptation. In practice, we should estimate this variance and update the two parameters online.

In the following, we present a simple numerical example to verify the theoretical conclusions and illustrate the improvements that may be achieved by optimizing the ϕ function. Consider the FIR system identification, where the transfer functions of the plant and the adaptive filter are [200]

$$\begin{cases} G^*(z) = 0.8 + 0.5z^{-1} \\ G(z) = w_0 + w_1 z^{-1} \end{cases} \tag{4.206}$$

The input signal and the noise are white Gaussian processes with powers 1.0 and 0.64, respectively. The initial weight vector of adaptive filter was obtained by perturbing each component of the optimal weight vector by a random variable that

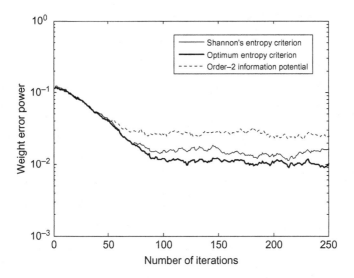

Figure 4.23 Convergence curves of three SIG algorithms with the same initial convergence rate
(adopted from Ref. [200]).

is uniformly distributed in the interval $[-0.6, 0.6]$. In the simulation, the SIG algorithms under three different entropy criteria (optimal entropy, Shannon entropy, QIP) are compared. The Gaussian kernel is used and the kernel size is kept fixed at $\sigma = 0.4$ during the adaptation.

First, the step-size of the SIG algorithm under the optimal entropy criterion was chosen to be $\eta = 0.015$. The step-sizes of the other two SIG algorithms are adjusted such that the three algorithms converge at the same initial rate. Figure 4.23 shows the average convergence curves over 300 simulation runs. Clearly, the optimal entropy criterion achieves the smallest final misadjustment (steady-state WEP). The step-sizes of the other two SIG algorithms can also be adjusted such that the three algorithms yield the same final misadjustment. The corresponding results are presented in Figure 4.24, which indicates that, beginning at the same initial WEP, the algorithm under optimal entropy criterion converges faster to the optimal solution than the other two algorithms. Therefore, a noticeable performance improvement can be achieved by optimizing the ϕ function.

Further, we consider the slow adaptation case in which the step-size for the optimal entropy criterion was chosen to be $\eta = 0.003$ (smaller than 0.015). It has been proved that, if the step-size becomes smaller (tends to zero), the optimal entropy will approach the Shannon entropy. Thus, in this case, the adaptation behavior of the SIG algorithm under Shannon entropy criterion would be nearly equivalent to that of the optimal entropy criterion. This theoretical prediction is confirmed by Figure 4.25, which illustrates that the Shannon entropy criterion and the optimal entropy criterion may produce almost the same convergence performance. In this figure, the initial convergence rates of the three algorithms are set equal.

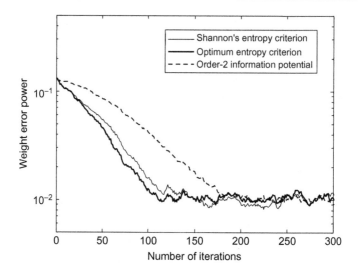

Figure 4.24 Convergence curves of three SIG algorithms with the same final misadjustment (adopted from Ref. [200]).

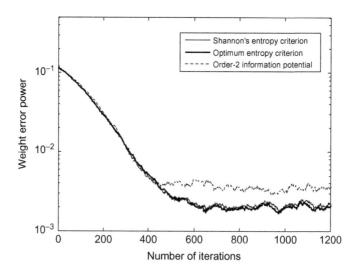

Figure 4.25 Convergence curves of three SIG algorithms in slow adaptation (adopted from Ref. [200]).

4.6   Survival Information Potential Criterion

Traditional entropy measures, such as Shannon and Renyi's entropies of a continuous random variable, are defined based on the PDF. As argued by Rao et al. [157], this kind of entropy definition has several drawbacks: (i) the definition will be ill-suited for the case in which the PDF does not exist; (ii) the value can be negative; and (iii) the approximation using empirical distribution is impossible in general. In order to overcome these problems, Rao et al. proposed a new definition of entropy based on the cumulative distribution function (or equivalently, the survival function) of a random variable, which they called the cumulative residual entropy (CRE) [157]. Motivated by the definition of CRE, Zografos and Nadarajah proposed two new broad classes of entropy measures based on the survival function, that is, the survival exponential and generalized survival exponential entropies, which include the CRE as a special case [158].

In the following, a new IP, namely the survival information potential (SIP) [159] is defined in terms of the survival function instead of the density function. The basic idea of this definition is to replace the density function with the survival function in the expression of the traditional IP. The SIP is, in fact, the argument of the power function in the survival exponential entropy. In a sense, this parallels the relationship between the IP and the Renyi entropy. When used as an adaptation criterion, the SIP has some advantages over the IP: (i) it has consistent definition in the continuous and discrete domains; (ii) it is not shift-invariant (i.e., its value will vary with the location of distribution); (iii) it can be easily computed from sample data (without kernel computation and the choice of kernel width), and the estimation asymptotically converges to the true value; and (iv) it is a more robust measure since the distribution function is more regular than the density function (note that the density function is computed as the derivative of the distribution function).

4.6.1   Definition of SIP

Before proceeding, we review the definitions of the CRE and survival exponential entropy.

Let $X = (X_1, X_2, \ldots, X_m)$ be a random vector in $\mathbb{R}^m$. Denote $|X|$ the absolute value transformed random vector of X, which is an m-dimensional random vector with components $|X_1|, |X_2|, \ldots, |X_m|$. Then the CRE of X is defined by [157]

$$\varepsilon(X) = -\int_{\mathbb{R}^m_+} \overline{F}_{|X|}(x) \log \overline{F}_{|X|}(x) \mathrm{d}x \qquad (4.207)$$

where $\overline{F}_{|X|}(x) = P(|X| > x) = E[I(|X| > x)]$ is the multivariate survival function of the random vector $|X|$, and $\mathbb{R}^m_+ = \{x \in \mathbb{R}^m : x = (x_1, \ldots, x_m), x_i \geq 0, i = 1, \ldots, m\}$. Here the notation $|X| > x$ means that $|X_i| > x_i$, $i = 1, \ldots, m$, and $I(.)$ denotes the indicator function.

Based on the same notations, the survival exponential entropy of order α is defined as [158]

$$M_\alpha(X) = \left(\int_{\mathbb{R}_+^m} \overline{F}_{|X|}^\alpha(x)\mathrm{d}x \right)^{1/(1-\alpha)} \tag{4.208}$$

From Eq. (4.208), we have

$$\log M_\alpha(X) = \frac{1}{1-\alpha}\log\int_{\mathbb{R}_+^m} \overline{F}_{|X|}^\alpha(x)\mathrm{d}x \tag{4.209}$$

It can be shown that the following limit holds [158]:

$$\lim_{\alpha\to 1}\left\{\log M_\alpha(X) - \frac{1}{1-\alpha}\log\int_{\mathbb{R}_+^m} \overline{F}_{|X|}(x)\mathrm{d}x\right\}\int_{\mathbb{R}_+^m} \overline{F}_{|X|}(x)\mathrm{d}x = \varepsilon(X) \tag{4.210}$$

The definition of the IP, along with the similarity between the survival exponential entropy and the Renyi entropy, motivates us to define the SIP.

Definition For a random vector X in $\mathbb{R}^m$, the SIP of order $\alpha(\alpha > 0)$ is defined by [159]

$$S_\alpha(X) = \int_{\mathbb{R}_+^m} \overline{F}_{|X|}^\alpha(x)\mathrm{d}x \tag{4.211}$$

The SIP (4.211) is just defined by replacing the density function with the survival function (of an absolute value transformation of X) in the original IP. This new definition seems more natural and reasonable, because the survival function (or equivalently, the distribution function) is more regular and general than the PDF. For the case $\alpha = 2$, we call the SIP the quadratic survival information potential (QSIP).

The SIP $S_\alpha(X)$ can be interpreted as the α-power of the α-norm in the survival functional space. When $\alpha < 1$, the survival exponential entropy $M_\alpha(X)$ is a monotonically increasing function of $S_\alpha(X)$, and minimizing the SIP is equivalent to minimizing the survival exponential entropy; while when $\alpha > 1$, the survival exponential entropy $M_\alpha(X)$ is a monotonically decreasing function of $S_\alpha(X)$, and in this case, minimizing the SIP is equivalent to maximizing the survival exponential entropy. We stress that when used as an approximation criterion in system identification, no matter what value of $\alpha > 0$, the SIP should be minimized to achieve smaller errors. This is quite different from the IP criterion which should be maximized when $\alpha > 1$ [64]. The reason for this is that $\forall \alpha > 0$, the smaller SIP

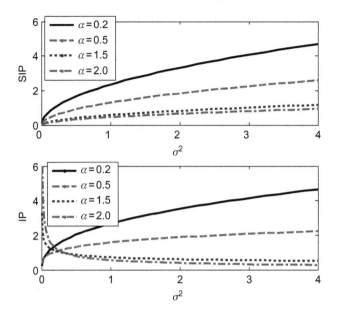

Figure 4.26 The SIP and IP for different σ^2 and α
(adopted from Ref. [159]).

corresponds to more concentrated errors around the zero value. To demonstrate this
fact, we give a simple example below.

Assume that X is zero-mean Gaussian distributed, $X \sim \mathcal{N}(0, \sigma^2)$. For different
variance σ^2 and α values, we can calculate the SIP and IP, which are shown in
Figure 4.26. It is clear that the SIP is a monotonically increasing function of σ^2 for
all the α values, while the IP is a monotonically increasing function only when
$\alpha < 1$.

4.6.2 Properties of the SIP

To further understand the SIP, we present in the following some important
properties.

Property 1: $\forall \alpha > 0$, $S_\alpha(X) = S_\alpha(|X|)$.

Proof: This property is a direct consequence of the definition of SIP.

Property 2: $S_\alpha(X) \geq 0$, with equality if and only if $P(X = 0) = 1$.

Proof: It is obvious that $S_\alpha(X) \geq 0$. Now $x^\alpha = 0$ if and only if $x = 0$. Therefore, $S_\alpha(X) = 0$ implies $P(|X| > \lambda) = 0$ for almost all $\lambda \in \mathbb{R}^m_+$, or in other word, for almost all $\lambda \in \mathbb{R}^m_+$, $P(|X| \leq \lambda) = 1$, which implies $P(X = \mathbf{0}) = 1$.

Remark: The global minimum value of the SIP is zero, and it corresponds to the δ distribution located at zero. This is a desirable property that fails to hold for conventional MEE criteria whose global minimum corresponds to the δ distribution located at any position (shift-invariant). Hence when using SIP as an approximation criterion in system identification, we do not need to add a bias term at system output.

The next five properties (Properties 3−7) are direct consequences of Ref. [158], and will, therefore, not be proved here (for detailed proofs, please refer to Theorems 1−5 in Ref. [158]).

Property 3: If $E[|X_i|] < \infty$ and $E[|X_i|^p] < \infty$ $(i = 1, 2, \ldots, m)$ for some $p > (m/\alpha)$, then $S_\alpha(X) < \infty$.

Property 4: Let X be an m-dimensional random vector, and let $Y = (Y_1, Y_2, \ldots, Y_m)$ with $Y_i = c_i X_i$, $c_i \in \mathbb{R}$, $i = 1, \ldots, m$. Then $S_\alpha(Y) = \left(\prod_{i=1}^m |c_i| \right) S_\alpha(X)$.

Property 5 (Weak convergence): Let $\{X(n)\}$ be a sequence of m-dimensional random vectors converging in law to a random vector X. If $\{X(n)\}$ are all bounded in L^p for some $p > m/\alpha$, then $\lim_{n \to \infty} S_\alpha(X(n)) = S_\alpha(X)$.

Property 6: If the components of an m-dimensional random vector X are independent with each other, then $S_\alpha(X) = \prod_{i=1}^m S_\alpha(X_i)$.

Property 7: Let X and Y be nonnegative and independent random variables $(X, Y \in \mathbb{R}_+)$. Then $S_\alpha(X + Y) \geq \max(S_\alpha(X), S_\alpha(Y))$.

Property 8: Given two m-dimensional continuous random vectors X and Y with PDFs $p_X(x)$ and $p_Y(y)$, if $p_X(x)$ is symmetric (rotation invariant for $m > 1$) and unimodal around zero, and Y is independent of X, then $S_\alpha(X + Y) \geq S_\alpha(X)$.

Proof: Since X and Y are independent,

$$p_{X+Y}(x) = \int_{\mathbb{R}^m} p_X(x - \tau) p_Y(\tau) d\tau \qquad (4.212)$$

It follows that $\forall \lambda \in \mathbb{R}^m_+$

$$
\begin{aligned}
\overline{F}_{|X+Y|}(\lambda) &= P(|X+Y| > \lambda) \\
&= \int_{\mathbb{R}^m} I(|x| > \lambda) p_{X+Y}(x) \mathrm{d}x \\
&= \int_{\mathbb{R}^m} I(|x| > \lambda) \mathrm{d}x \int_{\mathbb{R}^m} p_X(x-\tau) p_Y(\tau) \mathrm{d}\tau \\
&= \int_{\mathbb{R}^m} p_Y(\tau) \mathrm{d}\tau \int_{\mathbb{R}^m} I(|x| > \lambda) p_X(x-\tau) \mathrm{d}x \\
&= \int_{\mathbb{R}^m} p_Y(\tau) \mathrm{d}\tau \left(1 - \int_{\mathbb{R}^m} I(|x| \le \lambda) p_X(x-\tau) \mathrm{d}x\right) \qquad (4.213) \\
&= 1 - \int_{\mathbb{R}^m} p_Y(\tau) \mathrm{d}\tau \int_{\mathbb{R}^m} I(|x+\tau| \le \lambda) p_X(x) \mathrm{d}x \\
&\overset{(a)}{\ge} 1 - \int_{\mathbb{R}^m} p_Y(\tau) \mathrm{d}\tau \int_{\mathbb{R}^m} I(|x| \le \lambda) p_X(x) \mathrm{d}x \\
&= \int_{\mathbb{R}^m} I(|x| > \lambda) p_X(x) \mathrm{d}x \\
&= \overline{F}_{|X|}(\lambda)
\end{aligned}
$$

where (a) follows from the condition that $p_X(x)$ is symmetric (rotation invariance for $m > 1$) and unimodal around zero. Thus we get $S_\alpha(X + Y \ge S_\alpha(X)$.

Property 9: For the case $\alpha = 1$, the SIP of $X \in \mathbb{R}^m$ equals the expectation of $\prod_{i=1}^m |X_i|$.

Proof:

$$
\begin{aligned}
S_1(X) &= \int_{\mathbb{R}^m_+} \overline{F}_{|X|}(\tau) \mathrm{d}\tau \\
&= \int_{\mathbb{R}^m_+} E[I(|X| > \tau)] \mathrm{d}\tau \\
&= \int_{\mathbb{R}^m_+} E\left[\prod_{i=1}^m I(|X_i| > \tau_i)\right] \mathrm{d}\tau \qquad (4.214) \\
&= E\left[\int_{\mathbb{R}^m_+} \left(\prod_{i=1}^m I(|X_i| > \tau_i)\right) \mathrm{d}\tau\right] \\
&= E\left[\prod_{i=1}^m |X_i|\right]
\end{aligned}
$$

The above property can be generalized to the case where α is a natural number, as stated in Property 10.

Property 10: If $\alpha \in \mathcal{N}$, then $S_\alpha(X) = E\left[\prod_{i=1}^m |Z_i|\right]$, where $Z_i = \min\left(|X_i|, \left|Y_i^{(1)}\right|, \cdots, \left|Y_i^{(\alpha-1)}\right|\right)$, $\{Y^{(j)}\}_{j=1}^{\alpha-1}$ are independent and identically distributed (i.i.d.) random vectors which are independent of but have the same distribution with X.

Proof: As random vectors $\{Y^{(j)}\}$ are i.i.d., independent of but have the same distribution with X, we can derive

$$\overline{F}_{|Z|}(\tau) = E[I(|Z| > \tau)]$$

$$= E\left[\prod_{i=1}^m I(|Z_i| > \tau_i)\right]$$

$$= E\left[\prod_{i=1}^m I\left(\min\left(|X_i|, \left|Y_i^{(1)}\right|, \cdots, \left|Y_i^{(\alpha-1)}\right|\right) > \tau_i\right)\right]$$

$$= E\left[\prod_{i=1}^m \left(I(|X_i| > \tau_i)\prod_{j=1}^{\alpha-1} I\left(\left|Y_i^{(j)}\right| > \tau_i\right)\right)\right]$$

$$= E\left[\prod_{i=1}^m (I(|X_i| > \tau_i))\prod_{j=1}^{\alpha-1}\left(\prod_{i=1}^m I\left(\left|Y_i^{(j)}\right| > \tau_i\right)\right)\right] \qquad (4.215)$$

$$= E\left[I(|X| > \tau)\prod_{j=1}^{\alpha-1}\left(I\left(\left|Y^{(j)}\right| > \tau\right)\right)\right]$$

$$= E[I(|X| > \tau)]\prod_{j=1}^{\alpha-1} E\left[I\left(\left|Y^{(j)}\right| > \tau\right)\right]$$

$$= \prod_{j=1}^{\alpha} E[I(|X| > \tau)]$$

$$= \overline{F}_{|X|}^\alpha(\tau)$$

And hence

$$S_\alpha(X) = \int_{\mathbb{R}_+^m} \overline{F}_{|X|}^\alpha(\tau)\mathrm{d}\tau = \int_{\mathbb{R}_+^m} \overline{F}_{|Z|}(\tau)\mathrm{d}\tau \overset{\text{Property 9}}{=} E\left[\prod_{i=1}^m |Z_i|\right] \qquad (4.216)$$

The next property establishes a relationship between SIP and IP (which exists when X has PDF). A similar relationship has been proved by Rao et al. for their CRE (see Proposition 4 in [157]).

Property 11: Let X be a nonnegative random variable with continuous distribution. Then there exists a function ϕ such that the α-order IP of $Y = \phi(X)$ is related to $S_\alpha(X)$ via

$$V_\alpha(Y) = \frac{S_\alpha(X)}{(E[X])^\alpha} \tag{4.217}$$

Proof: Let $F(x)$ be the distribution function with density $P(X > x)/E[X]$. If we choose $\phi(x) = F^{-1}(F_X(x))$, where $F^{-1}(.)$ is defined as in the remarks preceding the Proposition 4 in [157], then $Y = \phi(X)$ has the distribution $F(x)$. Therefore, we have

$$V_\alpha(Y) = \int_{-\infty}^{\infty} p_Y^\alpha(\tau)d\tau = \int_{-\infty}^{\infty} \left(\frac{P(X > \tau)}{E[X]}\right)^\alpha d\tau = \frac{S_\alpha(X)}{(E[X])^\alpha} \tag{4.218}$$

Property 12: Let $X \in \mathbb{R}^m$ and $Y \in \mathbb{R}^n$ be two continuous random vectors. Assuming for every value $Y = y$, the conditional density $p_{X|Y}(x|y)$ is symmetric (rotation invariant for $m > 1$) and unimodal in x around $\mu(y) = E[X|Y = y]$, then $S_\alpha(X - \mu(Y)) \leq S_\alpha(X - g(Y))$, where $g(.)$ is any mapping $\mathbb{R}^n \rightarrow \mathbb{R}^m$ for which $S_\alpha(X - g(Y))$ exists.

Proof: Denote $p^\mu(x)$ and $p^g(x)$, respectively, the densities of $X - \mu(Y)$ and $X - g(Y)$, i.e.,

$$\begin{cases} p^\mu(x) = \int_{\mathbb{R}^n} p_{X|Y}(x + \mu(y)|y)dF_Y(y) \\ p^g(x) = \int_{\mathbb{R}^n} p_{X|Y}(x + g(y)|y)dF_Y(y) \end{cases} \tag{4.219}$$

Then for any $x \in \mathbb{R}_+^m$, we have

$$\begin{aligned}
\overline{F}_{|X-g(Y)|}(x) &= E[I(|X - g(Y)| > x)] \\
&= 1 - E[I(|X - g(Y)| \leq x)] \\
&= 1 - \int_{\mathbb{R}^m} I(|\tau| \leq x)p^g(\tau)d\tau \\
&= 1 - \int_{\mathbb{R}^m} I(|\tau| \leq x)d\tau \int_{\mathbb{R}^n} p_{X|Y}(\tau + g(y)|y)dF_Y(y) \\
&= 1 - \int_{\mathbb{R}^n} dF_Y(y) \int_{\mathbb{R}^m} I(|\tau| \leq x)p_{X|Y}(\tau + g(y)|y)d\tau \\
&= 1 - \int_{\mathbb{R}^n} dF_Y(y) \int_{\mathbb{R}^m} I(|\tau - g(y)| \leq x)p_{X|Y}(\tau|y)d\tau \\
&\overset{(b)}{\geq} 1 - \int_{\mathbb{R}^n} dF_Y(y) \int_{\mathbb{R}^m} I(||\tau - \mu(y)| \leq x)p_{X|Y}(\tau|y)d\tau \\
&= 1 - \int_{\mathbb{R}^m} I(|\tau| \leq x)p^\mu(\tau)d\tau \\
&= \overline{F}_{|X-\mu(Y)|}(x)
\end{aligned} \tag{4.220}$$

where (b) comes from the condition that for every y, $p_{X|Y}(x|y)$ is symmetric (rotation invariant for $m > 1$) and unimodal in x around $\mu(y)$. Therefore

$$S_\alpha(X - \mu(Y)) = \int_{\mathbb{R}^m_+} \overline{F}^\alpha_{|X-\mu(Y)|}(x)dx \leq \int_{\mathbb{R}^m_+} \overline{F}^\alpha_{|X-g(Y)|}(x)dx = S_\alpha(X - g(Y)) \qquad (4.221)$$

Remark: The above property suggests that under certain conditions, the conditional mean $\mu(Y)$, which minimizes the MSE, also minimizes the error's SIP.

4.6.3 Empirical SIP

Next, we discuss the empirical SIP of X. Since $S_\alpha(X) = S_\alpha(|X|)$ (see Property 1), we assume without loss of generality that $X \in \mathbb{R}^m_+$. Let $X(1), X(2), \ldots, X(N)$ be N i.i.d. samples of X with survival function $\overline{F}_X(x)$. The empirical survival function of X can be estimated by putting $1/N$ at each of the sample points, i.e.,

$$\overline{F}_N(x) = \frac{1}{N}\sum_{k=1}^N I(X(k) > x) \qquad (4.222)$$

Consequently, the empirical SIP can be calculated as

$$\hat{S}_\alpha(X) = \int_{\mathbb{R}^m_+} \overline{F}^\alpha_N(x)dx = \int_{\mathbb{R}^m_+} \left(\frac{1}{N}\sum_{k=1}^N I(X(k) > x)\right)^\alpha dx \qquad (4.223)$$

According to Glivento$-$Cantelli theorem [202],

$$\left\|\overline{F}_N - \overline{F}_X\right\|_\infty = \sup_x \left|\overline{F}_N(x) - \overline{F}_X(x)\right| \xrightarrow[N \to \infty]{a.s.} 0 \qquad (4.224)$$

Combining Eq. (4.224) and Property 5 yields the following proposition.

Proposition: For any random vector X in $\mathbb{R}^m_+$, if X is bounded in L^p for some $P > m/\alpha$, then the empirical SIP (4.223) will converge to the true SIP of X, i.e., $\lim_{N \to \infty} \hat{S}_\alpha(X) = S_\alpha(X)$.

In the sequel, we will derive more explicit expressions for the empirical SIP (4.223). Let $x(1), x(2), \ldots, x(N)$ be a realization of $X(1), X(2), \ldots, X(N)$.

4.6.3.1 Scalar Data Case

First, we consider the scalar data case, i.e., $m = 1$. We assume, without loss of generality, that $0 \leq x(1) \leq x(2) \leq \cdots \leq x(N)$. Then we derive

$$\hat{S}_\alpha(X) = \int_0^\infty \left(\frac{1}{N} \sum_{k=1}^N I(x(k) > x) \right)^\alpha dx$$

$$= \sum_{j=1}^N \int_{x(j-1)}^{x(j)} \left(\frac{1}{N} \sum_{k=1}^N I(x(k) > x) \right)^\alpha dx \qquad (4.225)$$

$$= \sum_{j=1}^N \left(\frac{N-j+1}{N} \right)^\alpha (x(j) - x(j-1))$$

where we assume $x(0) = 0$. One can rewrite Eq. (4.225) into a more simple form:

$$\hat{S}_\alpha(X) = \sum_{j=1}^N \left(\frac{N-j+1}{N} \right)^\alpha (x(j) - x(j-1))$$

$$= \left(1 - \left(\frac{N-1}{N} \right)^\alpha \right) x(1) + \left(\left(\frac{N-1}{N} \right)^\alpha - \left(\frac{N-2}{N} \right)^\alpha \right) x(2) + \cdots +$$

$$\left(\left(\frac{2}{N} \right)^\alpha - \left(\frac{1}{N} \right)^\alpha \right) x(N-1) + \left(\frac{1}{N} \right)^\alpha x(N) \qquad (4.226)$$

$$= \sum_{j=1}^N \lambda_j x(j)$$

where

$$\lambda_j = \left(\frac{N-j+1}{N} \right)^\alpha - \left(\frac{N-j}{N} \right)^\alpha \qquad (4.227)$$

From Eq. (4.226), the empirical SIP for scalar data can be expressed as a weighted sum of the ordered sample data $0 \le x(1) \le x(2) \le \cdots \le x(N)$, where the weights $\lambda_j, j = 1, 2, \ldots, N$ depend on the sample size N and the α value, satisfying $\lambda_j \ge 0$, $\sum_{j=1}^N \lambda_j = 1$. For the case $N = 10$, the weights for different α values are shown in Figure 4.27. One can observe: (i) when $\alpha = 1.0$, all the weights are equal ($\lambda_j = 1/N$), and in this case the empirical SIP is identical to the sample mean of X and (ii) when $\alpha \ne 1.0$, the weights are not equal. Specifically, when $\alpha < 1.0$ ($\alpha > 1.0$), the weight λ_j is a monotonically increasing (decreasing) function of the order index j, that is, the larger weights are assigned to the larger (smaller) sample data.

4.6.3.2 Multidimensional Data Case

Computing the empirical SIP for multidimensional data ($m > 1$) is, in general, not an easy task. If α is a natural number, however, we can still obtain a simple explicit expression. In this case, one can derive

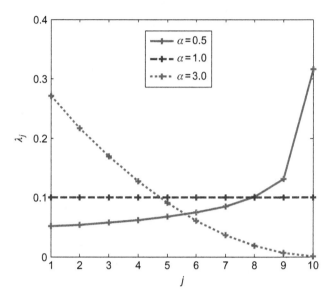

Figure 4.27 The weights for different α values
(adopted from Ref. [159]).

$$
\begin{aligned}
\hat{S}_\alpha(X) &= \int_{\mathbb{R}_+^m} \left(\frac{1}{N} \sum_{k=1}^N I(x(k) > x) \right)^\alpha dx \\
&= \frac{1}{N^\alpha} \int_{\mathbb{R}_+^m} \sum_{j_1, \cdots, j_\alpha = 1}^N (I(x(j_1) > x) \times \cdots \times I(x(j_\alpha) > x)) dx \\
&= \frac{1}{N^\alpha} \sum_{j_1, \cdots, j_\alpha = 1}^N \int_{\mathbb{R}_+^m} (I(x(j_1) > x) \times \cdots \times I(x(j_\alpha) > x)) dx \\
&= \frac{1}{N^\alpha} \sum_{j_1, \cdots, j_\alpha = 1}^N \int_{\mathbb{R}_+^m} \left(\prod_{i=1}^m I(\min(x_i(j_1), \ldots, x_i(j_\alpha)) > x_i) \right) dx \\
&= \frac{1}{N^\alpha} \sum_{j_1, \cdots, j_\alpha = 1}^N \left(\prod_{i=1}^m \int_{\mathbb{R}_+} I(\min(x_i(j_1), \ldots, x_i(j_\alpha)) > x_i) dx_i \right) \\
&= \frac{1}{N^\alpha} \sum_{j_1, \cdots, j_\alpha = 1}^N \left(\prod_{i=1}^m \min(x_i(j_1), \ldots, x_i(j_\alpha)) \right)
\end{aligned}
\tag{4.228}
$$

The empirical SIP in Eq. (4.228) can also be derived using Property 10. By Property 10, we have $S_\alpha(X) = E\left[\prod_{i=1}^m \min(X_i, Y_i^{(1)}, \ldots, Y_i^{(\alpha-1)}) \right]$, where $\{Y^{(j)}\}$ are i.i.d. random vectors which are independent of but have the same distribution with

X. It is now evident that the empirical SIP of Eq. (4.228) is actually the sample mean estimate of $E\left[\prod_{i=1}^{m}\min(X_i, Y_i^{(1)}, \ldots, Y_i^{(\alpha-1)})\right]$.

Remark: Compared with the empirical IP (say the estimated IP by Parzen window method), the empirical SIP is much simpler in computation (just an ordering of the samples), since there is no kernel evaluation, the most time-consuming part in the calculation of the empirical IP. In addition, there is no problem like kernel width choice.

4.6.4 Application to System Identification

Similar to the IP, the SIP can also be used as an optimality criterion in adaptive system training. Under the SIP criterion, the unknown system parameter vector (or weight vector) W can be estimated as

$$\hat{W} = \arg\min_{W \in \Omega_W} S_\alpha(e_k) = \arg\min_{W \in \Omega_W} \int_{R_+^m} \overline{F}_{|e_k|}^\alpha(\xi)d\xi \qquad (4.229)$$

where $\overline{F}_{|e_k|}(.)$ is the survival function of the absolute value transformed error $|e_k| = |z_k - \hat{y}_k|$. In practical application, the error distribution is usually unknown; we have to use, instead of the theoretical SIP, the empirical SIP as the cost function. Given a sequence of error samples $(e_1, e_2, \ldots, e_N)$, assuming, without loss of generality, that $|e_1| \leq |e_2| \leq \cdots \leq |e_N|$, the empirical SIP will be (assume scalar error)

$$\hat{S}_\alpha(e) = \sum_{j=1}^{N} \lambda_j |e_j| \qquad (4.230)$$

where λ_j is calculated by Eq. (4.227). The empirical cost (4.230) is a weighted sum of the ordered absolute errors. One drawback of Eq. (4.230) is that it is not smooth at $e_j = 0$. To address this problem, one can use the empirical SIP of the square errors $(e_1^2, e_2^2, \ldots, e_N^2)$ as an alternative adaptation cost, given by

$$\hat{S}_\alpha(e^2) = \sum_{j=1}^{N} \lambda_j e_j^2 \qquad (4.231)$$

The above cost is the weighted sum of the ordered square errors, which includes the popular MSE cost as a special case (when $\alpha = 1$). A more general cost can be defined as the empirical SIP of any mapped errors $(\phi(e_1), \phi(e_2), \ldots, \phi(e_N))$, i.e.,

$$\hat{S}_\alpha(\phi(e)) = \sum_{j=1}^{N} \lambda_j \phi(e_j) \qquad (4.232)$$

where function $\phi(.)$ usually satisfies

$$\begin{cases} (i) \text{ positivity:} & \phi(e) \geq 0 \\ (ii) \text{ symmetry:} & \phi(e) = \phi(-e) \\ (iii) \text{ monotonicity:} & |e_1| < |e_2| \Rightarrow \phi(e_1) \leq \phi(e_2) \end{cases} \tag{4.233}$$

Based on the general cost (4.232), the weight update equation for system identification is

$$W_{k+1} = W_k - \eta \sum_{j=1}^{N} \lambda_j \phi'(e_j) \partial e_j / \partial W \tag{4.234}$$

The weight update can be performed online (i.e., over a short sliding window), as described in Table 4.5.

In the following, we present two simulation examples to demonstrate the performance of the SIP minimization criterion. In the simulations below, the empirical cost (4.231) is adopted ($\phi(e) = e^2$).

Table 4.5 Online System Identification with SIP Criterion

Initialization
a. Initialize the weight vector of the adaptive system: W_0
b. Choose the α value, step-size η, and the sliding window length L
c. Compute the weights $\lambda_j, j = 1, \ldots, L$, using Eq. (4.227)
d. Initialize the window of errors: $(e(1), \ldots, e(L)) = (0, \cdots, 0)$
Computation
  while$\{x_k, z_k\}$ available **do**

  1. Compute the error: $e_k = z_k - \hat{y}_k$
  2. Update the window of errors:

$$\begin{cases} e(j) = e(j+1), & \text{for } j = 1, \ldots, L-1 \\ e(L) = e_k \end{cases}$$

  3. Rearrange the errors in ascending order of magnitude:

$$|e(1)| \leq |e(2)| \leq \cdots \leq |e(L)|$$

  4. Update the weight vector:

$$W_{k+1} = W_k - \eta \sum_{j=1}^{L} \lambda_j \phi'(e(j)) \partial e(j) / \partial W$$

  end while

4.6.4.1 FIR System Identification

First, we consider the simple FIR system identification. Let the unknown system be a FIR filter given by [159]

$$H(z) = 0.1 + 0.2z^{-1} + 0.3z^{-2} + 0.4z^{-3} + 0.5z^{-4} + 0.6z^{-5}$$
$$+ 0.5z^{-6} + 0.4z^{-7} + 0.3z^{-8} + 0.2z^{-9} + 0.1z^{-10} \tag{4.235}$$

The adaptive system is another FIR filter with the same order. The input x_k is a white Gaussian process with unit variance. Assume that the output of unknown system is disturbed by an additive noise. Three different distributions are utilized to generate the noise data:

$$\begin{cases} (a)\,\text{Symmetric } \alpha\text{-stable } (S\alpha S)\!: \quad \psi_{\gamma,\alpha}(\omega) = \exp(-\gamma|\omega|^{\alpha}) \text{ with } \gamma = 0.1, \alpha = 1.5 \\ (b)\,\text{Gaussian:} \quad p(x) = \dfrac{1}{\sqrt{2\pi}\sigma} \exp(-x^2/2\sigma^2) \text{ with } \sigma^2 = 0.2 \\ (c)\,\text{Binary:} \quad \Pr(x=0.5) = 0.5, \quad \Pr(x=-0.5) = 0.5 \end{cases}$$
$$\tag{4.236}$$

where $\psi_{\gamma,\alpha}(\omega)$ denotes the characteristic function of the $S\alpha S$ distribution. The above three distributions have, respectively, heavy, medium, and light tails. The noise signals are shown in Figure 4.28.

In the simulation, the sliding data length is set at 10. The step-sizes of each algorithm are chosen such that the initial convergence rates are visually identical. Figure 4.29 shows the average convergence curves over 100 Monte Carlo runs for different $\alpha\,(0.5, 1.0, 2.0)$. Notice that when $\alpha = 1.0$, the algorithm is actually the

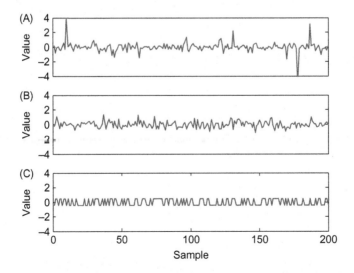

Figure 4.28 Three different noises: (A) $S\alpha S$, (B) Gaussian, and (C) Binary.

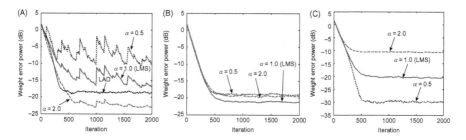

Figure 4.29 Convergence curves averaged over 100 Monte Carlo runs: (A) $S\alpha S$, (B) Gaussian, and (C) Binary.

Table 4.6 WEPs at Final Iteration Over 100 Monte Carlo Runs

	$S\alpha S$	**Gaussian**	**Binary**
$\alpha = 0.5$	0.0989 ± 0.1700	0.0127 ± 0.0055	0.0010 ± 0.0005
$\alpha = 1.0$	0.0204 ± 0.0294	0.0076 ± 0.0029	0.0096 ± 0.0039
$\alpha = 2.0$	0.0051 ± 0.0029	0.0112 ± 0.0043	0.0880 ± 0.0133

LMS algorithm (strictly speaking, the block LMS algorithm). The WEPs at final iteration are summarized in Table 4.6. From simulation results, one can observe:

i. For the case of SαS noise, the algorithms with larger α values (say $\alpha = 2.0$) converge to smaller WEP, and can even outperform the LAD algorithm (for comparison purpose, we also plot in Figure 4.29(A) the convergence curve of LAD). It is well known that the LAD algorithm performs well in α-stable noises [31].

ii. For the case of Gaussian noise, the algorithm with $\alpha = 1.0$ (the LMS algorithm) performs better, which is to be expected, since MSE criterion is optimal for linear Gaussian systems.

iii. For the case of binary noise, the algorithms with smaller α values (say $\alpha = 0.5$) obtain better performance.

The basic reasons for these findings are as follows. As shown in Figure 4.27, the larger the α value, the smaller the weights assigned to the larger errors. For the case of heavy-tail noises (e.g., SαS noise), the larger errors are usually caused by the impulsive noise. In this case, the larger α value will reduce the influence of the outliers and improve the performance. On the other hand, for the case of light-tail noises (e.g., binary noise), the larger errors are mainly caused by the system mismatch, thus the smaller α value will decrease the larger mismatch more rapidly (as the larger weights are assigned to the larger errors).

4.6.4.2   TDNN Training

The second simulation example is on the TDNNs training (in batch mode) with SIP minimization criterion for one-step prediction of the Mackey−Glass (MG) chaotic

Table 4.7 Testing Errors Over 4000 Test Samples in MG Time Series Prediction

	SIP		IP
$\alpha = 0.8$	0.0016 ± 0.0276	$\alpha = 1.05$	0.0011 ± 0.0310
$\alpha = 1.0$	0.0011 ± 0.0214	$\alpha = 1.5$	0.0010 ± 0.0206
$\alpha = 1.5$	0.0002 ± 0.0180	$\alpha = 2.0$	0.0006 ± 0.0195
$\alpha = 2.0$	-0.0019 ± 0.0272	$\alpha = 2.5$	0.0009 ± 0.0203
$\alpha = 2.5$	-0.0085 ± 0.0412	$\alpha = 3.0$	0.0010 ± 0.0218

time series [203] with delay parameter $\tau = 30$ and sampling period 6 s. The TDNN is built from MLPs that consist of six processing elements (PEs) in a hidden layer with biases and tanh nonlinearities and a single linear output PE with an output bias. The goal is to predict the value of the current sample x_k using the previous seven points $X_k = \{x_{k-1}, \ldots, x_{k-7}\}$ (the size of the input delay line is consistent with Taken's embedding theorem [204]). In essence, the problem is to identify the underlying mapping between the input vector X_k and the desired output x_k. For comparison purpose, we also present simulation results of TDNN training with α-order ($\alpha > 1$) IP maximization criterion. Since IP is shift-invariant, after training the bias value of the output PE was adjusted so as to yield zero-mean error over the training set. The Gaussian kernel was used to evaluate the empirical IP and the kernel size was experimentally set at 0.8. A segment of 200 samples is used as the training data. To avoid local-optimal solutions, each TDNN is trained starting from 500 predetermined initial weights generated by zero-mean Gaussian distribution with variance 0.01. The best solution (the one with the lowest SIP or the highest IP after training) among the 500 candidates is selected to test the accuracy performance. In each simulation, the training algorithms utilized BP with variable stepsizes [205], and 1000 iterations were run to ensure the convergence. The trained networks are tested on an independently generated test sequence of 4000 samples, and the testing errors are listed in Table 4.7. One can see the TDNN trained using SIP with $\alpha = 1.5$ achieves the smallest testing error. Thus, if properly choosing the order α, the SIP criterion is capable of outperforming the IP criterion. Figure 4.30 shows the computation time per iteration versus the number of training data. Clearly, the SIP-based training is computationally much more efficient than the IP-based training, especially for large data sets. The training time for both methods is measured on a personal computer equipped with a 2.2 GHz Processor and 3 GB memory.

4.7 Δ-Entropy Criterion

System identification usually handles continuous-valued random processes rather than discrete-valued processes. In many practical situations, however, the input

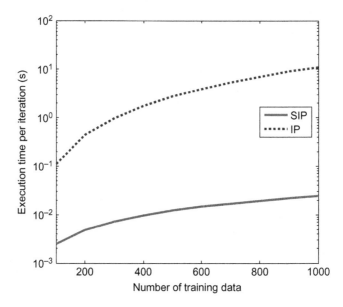

Figure 4.30 Execution time per iteration versus the number of training data.

and/or output of the unknown system may be discrete-valued for a variety of reasons:

a. For many systems, especially in the field of digital communication, the input signals take values only in finite alphabetical sets.

b. Coarsely quantized signals are commonly used when the data are obtained from an A/D converter or from a communication channel. Typical contexts involving quantized data include digital control systems (DCSs), networked control systems (NCSs), wireless sensor networks (WSNs), etc.

c. Binary-valued sensors occur frequently in practical systems. Some typical examples of binary-valued sensors can be found in Ref. [206].

d. Discrete-valued time series are common in practice. In recent years, the count or integer-valued data time series have gained increasing attentions [207−210].

e. Sometimes, due to computational consideration, even if the observed signals are continuous-valued, one may classify the data into groups and obtain the discrete-valued data [130, Chap. 5].

In these situations, one may apply the differential entropy (or IP) to implement the MEE criterion, in spite of the fact that the random variables are indeed discrete. When the discretization is coarse (i.e., few levels) the use of differential entropy may carry a penalty in performance that is normally not quantified. Alternatively, the MEE implemented with discrete entropy will become ill-suited since the minimization fails to constrain the dispersion of the error value which should be pursued because the error dynamic range decreases over iterations.

In the following, we augment the MEE criterion choices by providing a new entropy definition for discrete random variables, called the Δ-entropy, which comprises two terms: one is the discrete entropy and the other is the logarithm of the average interval between two successive discrete values. This new entropy retains important properties of the differential entropy and reduces to the traditional discrete entropy for a special case. More importantly, the proposed entropy definition can still be used to measure the value dispersion of a discrete random variable, and hence can be used as an MEE optimality criterion in system identification with discrete-valued data.

4.7.1 Definition of Δ-Entropy

Before giving the definition of Δ-entropy, let's review a fundamental relationship between the differential entropy and discrete entropy (for details, see also Ref. [43]).

Consider a continuous scalar random variable X with PDF $f(x)$. One can produce a quantized random variable X^Δ (see Figure 4.31), given by

$$X^\Delta = s_i, \quad \text{if} \quad i\Delta \leq X < (i+1)\Delta \tag{4.237}$$

where s_i is one of countable values, satisfying

$$i\Delta \leq s_i < (i+1)\Delta, \quad \text{and} \quad f(s_i)\Delta = \int_{i\Delta}^{(i+1)\Delta} f(x)\mathrm{d}x \tag{4.238}$$

The probability that $X^\Delta = s_i$ is

$$p_i = \Pr(X^\Delta = s_i) = f(s_i)\Delta \tag{4.239}$$

Figure 4.31 Quantization of a continuous random variable.

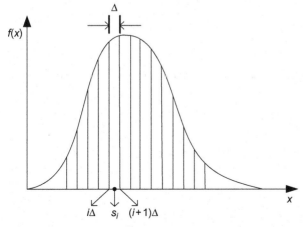

And hence, the discrete entropy $H(X^\Delta)$ is calculated as

$$H(X^\Delta) = - \sum_{i=-\infty}^{\infty} p_i \log p_i = - \sum_{i=-\infty}^{\infty} \Delta f(s_i) \log f(s_i) - \log \Delta \qquad (4.240)$$

If the density function $f(x)$ is Riemann integrable, the following limit holds

$$\lim_{\Delta \to 0} (H(X^\Delta) + \log \Delta) = - \int_{-\infty}^{\infty} f(x) \log f(x) dx = h(X) \qquad (4.241)$$

Here, to make a distinction between the discrete entropy and differential entropy, we use $h(X)$ instead of $H(X)$ to denote the differential entropy of X. Thus, if the quantization interval Δ is small enough, we have

$$h(X) \approx H(X^\Delta) + \log \Delta \qquad (4.242)$$

So, the differential entropy of a continuous random variable X is approximately equal to the discrete entropy of the quantized variable X^Δ plus the logarithm of the quantization interval Δ. The above relationship explains why differential entropy is sensitive to value dispersion. That is, compared with the discrete entropy, the differential entropy "contains" the term $\log \Delta$, which measures the average interval between two successive quantized values since

$$\Delta = \lim_{N \to \infty} \frac{1}{2N + 1} \sum_{i=-N}^{N} |s_{i+1} - s_i| \qquad (4.243)$$

This important relationship also inspired us to seek a new entropy definition for discrete random variables that will measure uncertainty as well as value dispersion and is defined as follows.

Definition: Given a discrete random variable X with values $\mathbf{S} = (s_1, s_2, \ldots, s_M)$, and the corresponding distribution $\mathbf{P} = (p_1, p_2, \ldots, p_M)$, the Δ-entropy, denoted by $H_\Delta(X)$ or $H_\Delta(\mathbf{S}, \mathbf{P})$, is defined as [211]

$$H_\Delta(X) = - \sum_{i=1}^{M} p_i \log p_i + \log \Delta(X) \qquad (4.244)$$

where $\Delta(X)$ (or $\Delta(\mathbf{S}, \mathbf{P})$) stands for the average interval (distance) between two successive values.

The Δ-entropy contains two terms: the first term is identical to the traditional discrete entropy and the second term equals the logarithm of the average interval between two successive values. This new entropy can be used as an optimality criterion in estimation or identification problems, because minimizing error's Δ-entropy decreases the average interval and automatically force the error values to concentrate.

Next, we discuss how to calculate the average interval $\Delta(X)$. Assume without loss of generality that the discrete values satisfy $s_1 < s_2 < \cdots < s_M$. Naturally, one immediately thinks of the arithmetic and geometric means, i.e.,

$$
\begin{cases}
\Delta(X) = \dfrac{1}{M-1} \sum_{i=1}^{M-1} |s_{i+1} - s_i| & \text{for arithmetic mean} \\[3mm]
\Delta(X) = \left(\prod_{i=1}^{M-1} |s_{i+1} - s_i| \right)^{1/(M-1)} & \text{for geometric mean}
\end{cases}
\tag{4.245}
$$

Both arithmetic and geometric means take no account of the distribution. A more reasonable approach is to calculate the average interval using a *probability-weighted method*. For example, one can use the following formula:

$$
\Delta(X) = \sum_{i=1}^{M-1} |s_{i+1} - s_i| \frac{p_i + p_{i+1}}{2}
\tag{4.246}
$$

However, if $(p_1 + p_M) > 0$, the sum of weights will be < 1, because

$$
\sum_{i=1}^{M-1} \frac{p_i + p_{i+1}}{2} = 1 - \frac{p_1 + p_M}{2} < 1
\tag{4.247}
$$

To address this issue, we give another formula:

$$
\Delta(X) = \sum_{i=1}^{M-1} |s_{i+1} - s_i| \frac{p_i + p_{i+1}}{2} + \frac{|s_M - s_1|}{M-1} \frac{p_1 + p_M}{2}
\tag{4.248}
$$

The second term of (4.248) equals the arithmetic mean multiplied by $(p_1 + p_M)/2$, which normalizes the weight sum to one. Substituting (4.248) into (4.244), we obtain

$$
H_\Delta(X) = - \sum_{i=1}^{M} p_i \log p_i + \log \left(\sum_{i=1}^{M-1} |s_{i+1} - s_i| \frac{p_i + p_{i+1}}{2} + \frac{|s_M - s_1|}{M-1} \frac{p_1 + p_M}{2} \right)
\tag{4.249}
$$

The Δ-entropy can be immediately extended to the infinite value-set case, i.e.,

$$
H_\Delta(X) = - \sum_{i=-\infty}^{\infty} p_i \log p_i + \log \left(\sum_{i=-\infty}^{\infty} |s_{i+1} - s_i| \frac{p_i + p_{i+1}}{2} + \lim_{N \to \infty} \frac{|s_N - s_{-N}|}{2N} \frac{p_{-N} + p_N}{2} \right)
\tag{4.250}
$$

In the following, we use Eq. (4.249) or (4.250) as the Δ-entropy expression.

4.7.2 Some Properties of the Δ-Entropy

The Δ-entropy maintains a close connection to the differential entropy. It is clear that the Δ-entropy and the differential entropy have the following relationship in the limit:

Theorem 1 For any continuous random variable X with Riemann integrable PDF $f(x)$, we have $\lim_{\Delta \to 0} H_\Delta(X^\Delta) = h(X)$, where quantized variable X^Δ is given by Eq. (4.237).

Proof: Combining Eqs. (4.239) and (4.250), we have

$$H_\Delta\left(X^\Delta\right) = -\sum_{i=-\infty}^{\infty} f(s_i)\Delta\log(f(s_i)\Delta)$$

$$+ \log\left(\sum_{i=-\infty}^{\infty} |s_{i+1} - s_i|\frac{f(s_i)\Delta + f(s_{i+1})\Delta}{2} + \lim_{N \to \infty}\frac{|s_N - s_{-N}|f(s_{-N})\Delta + f(s_N)\Delta}{2N}\frac{}{2}\right)$$

$$= -\sum_{i=-\infty}^{\infty} \Delta f(s_i)\log f(s_i) + \log\left(\sum_{i=-\infty}^{\infty} |s_{i+1} - s_i|\frac{f(s_i) + f(s_{i+1})}{2}\right)$$

As $f(x)$ is Riemann integrable, it follows that

$$\lim_{\Delta \to 0} H_\Delta(X^\Delta) = -\int_{-\infty}^{\infty} f(x)\log f(x)\mathrm{d}x + \log\left(\int_{-\infty}^{\infty} f(x)\mathrm{d}x\right)$$
$$= -\int_{-\infty}^{\infty} f(x)\log f(x)\mathrm{d}x = h(x)$$

This completes the proof.

Remark: The differential entropy of X is the limit of the Δ-entropy of X^Δ as $\Delta \to 0$. Thus, to some extent one can regard the Δ-entropy as a "quantized version" of the differential entropy.

Theorem 2 $\log\left(\max_{j=1,2,\cdots,M-1}|s_{j+1} - s_j|\right) \geq H_\Delta(X) - H(X) \geq \log\left(\min_{j=1,2,\cdots,M-1}|s_{j+1} - s_j|\right)$.

Proof: Omitted due to simplicity.

Remark: By Theorem 2, if the minimum interval between two successive values is larger than 1, we have $H_\Delta(X) > H(X)$, whereas if the maximum interval between two successive values is smaller than 1, we have $H_\Delta(X) < H(X)$.

Theorem 3 If X is a discrete random variable with equally spaced values, and the interval $\Delta = 1$, then $H_\Delta(X) = H(X)$.

Proof: For equally spaced intervals, the difference between the Δ-entropy and the discrete entropy equals $\log \Delta$. Hence, the statement follows directly.

Remark: The classification problem is, in general, a typical example of the error variable distributed on equally spaced values $\{0, 1, 2, 3, \ldots\}$. Thus in classification, the error's discrete entropy is equivalent to the Δ-entropy. This fact also gives an interpretation for why the discrete entropy can be used in the test and classification problems [212,213].

In information theory, it has been proved that the discrete entropy satisfies (see Ref. [43], p. 489)

$$0 \leq H(X) \leq \frac{1}{2}\log\left(2\pi e\left(\sum_{i=1}^{M} p_i i^2 - \left(\sum_{i=1}^{M} i p_i\right) + \frac{1}{12}\right)\right) \tag{4.251}$$

Combining Eq. (4.251) and Theorem 2, we obtain a bound on Δ-entropy:

$$\log\left(\min_{j=1,2,\cdots,M-1}\left|s_{j+1} - s_j\right|\right) \leq H_\Delta(X)$$
$$\leq \frac{1}{2}\log\left(2\pi e\left(\sum_{i=1}^{M} p_i i^2 - \left(\sum_{i=1}^{M} i p_i\right) + \frac{1}{12}\right)\left(\max_{j=1,2,\cdots,M-1}\left|s_{j+1} - s_j\right|\right)^2\right) \tag{4.252}$$

A lower bound of the Δ-entropy can also be expressed in terms of the variance $\mathrm{Var}(X)$, as given in the following theorem:

Theorem 4 If $p_{\min} = \min\{p_i\} > 0$, then $H_\Delta(X) \geq \log\left(\frac{2Mp_{\min}}{M-1}\right) + \frac{1}{2}\log(\mathrm{Var}(X))$.

Proof: It is easy to derive

$$\mathrm{Var}(X) = \sum_{i=1}^{M}(s_i - \bar{s})^2 p_i$$
$$\leq \sum_{i=1}^{M}\left(s_i - \frac{s_M + s_1}{2}\right)^2 p_i$$
$$\leq \sum_{i=1}^{M}\left(s_M - \frac{s_M + s_1}{2}\right)^2 p_i$$
$$= \frac{1}{4}(s_M - s_1)^2$$

It follows that $|s_M - s_1| \geq 2\sqrt{\text{Var}(X)}$, and hence

$$
H_\Delta(X) \geq \log\left(\sum_{i=1}^{M-1} |s_{i+1} - s_i| \frac{p_i + p_{i+1}}{2} + \frac{|s_M - s_1|}{M-1} \frac{p_1 + p_M}{2} \right)
$$

$$
\geq \log\left(\sum_{i=1}^{M-1} |s_{i+1} - s_i| p_{\min} + \frac{|s_M - s_1|}{M-1} p_{\min} \right) = \log\left(\frac{M|s_M - s_1|}{M-1} p_{\min} \right)
$$

$$
\geq \log\left(\frac{2M\sqrt{\text{Var}(X)}}{M-1} p_{\min} \right)
$$

$$
= \log\left(\frac{2Mp_{\min}}{M-1} \right) + \frac{1}{2}\log(\text{Var}(X))
$$

The lower bound of Theorem 4 suggests that, under certain condition minimizing the Δ-entropy constrains the variance. This is a key difference between the Δ-entropy and conventional discrete entropy.

Theorem 5 For any discrete random variable X, $\forall c \in \mathbb{R}$, $H_\Delta(X + c) = H_\Delta(X)$.

Proof: Since $H(X + c) = H(X)$ and $\Delta(X + c) = \Delta(X)$, we have $H_\Delta(X + c) = H_\Delta(X)$.

Theorem 6 $\forall \alpha \in \mathbb{R}$, $\alpha \neq 0$, $H_\Delta(\alpha X) = H_\Delta(X) + \log|\alpha|$.

Proof: Since $H(\alpha X) = H(X)$ and $\Delta(\alpha X) = |\alpha|\Delta(X)$,   we   have $H_\Delta(\alpha X) = H_\Delta(X) + \log|\alpha|$.

Theorems 5 and 6 indicate that the Δ-entropy has the same shifting and scaling properties as the differential entropy.

Theorem 7 The Δ-entropy is a concave function of $P = (p_1, p_2, \ldots, p_M)$.

Proof: $\forall P_1 = (p_1^{(1)}, p_2^{(1)}, \ldots, p_M^{(1)})$, $P_2 = (p_1^{(2)}, p_2^{(2)}, \ldots, p_M^{(2)})$, and $\forall 0 \leq \lambda \leq 1$, we have

$$
\Delta(S, \lambda P_1 + (1 - \lambda)P_2) = \lambda\Delta(S, P_1) + (1 - \lambda)\Delta(S, P_2) \tag{4.253}
$$

By the concavity of the logarithm function,

$$
\log(\Delta(S, \lambda P_1 + (1 - \lambda)P_2)) \geq \lambda \log(\Delta(S, P_1)) + (1 - \lambda)\log(\Delta(S, P_2)) \tag{4.254}
$$

It is well known that the discrete entropy $H(P)$ is a concave function of the distribution P, i.e.,

$$H(\lambda P_1 + (1 - \lambda)P_2) \geq \lambda H(P_1) + (1 - \lambda)H(P_2), \quad \forall\, 0 \leq \lambda \leq 1 \tag{4.255}$$

Combining Eqs. (4.254) and (4.255) yields

$$H_\Delta(S, \lambda P_1 + (1 - \lambda)P_2) \geq \lambda H_\Delta(S, P_1) + (1 - \lambda)H_\Delta(S, P_2) \tag{4.256}$$

which implies Δ-entropy is a concave function of P.

The concavity of the Δ-entropy is a desirable property for the entropy optimization problem. This property ensures that when a stationary value of the Δ-entropy subject to linear constraints is found, it gives the global maximum value [149].

Next, we solve the maximum Δ-entropy distribution. Consider the following constrained optimization problem:

$$\begin{cases} \max_{P} H_\Delta(X) \\ \text{s.t.} \begin{cases} \sum_{i=1}^{M} p_i = 1 \\ \sum_{i=1}^{M} p_i g_k(s_i) = a_k, \quad k = 1, 2, \ldots, K \end{cases} \end{cases} \tag{4.257}$$

where a_k is the expected value of the function $g_k(X)$. The Lagrangian is given by

$$L = H_\Delta(X) - (\lambda_0 - 1)\left(\sum_{i=1}^{M} p_i - 1\right) - \sum_{k=1}^{K} \lambda_k \left(\sum_{i=1}^{M} p_i g_k(s_i) - a_k\right) \tag{4.258}$$

where $\lambda_0, \lambda_1, \ldots, \lambda_K$ are the $(K + 1)$ Lagrange multipliers corresponding to the $(K + 1)$ constraints. Here $\lambda_0 - 1$ is used as the first Lagrange multiplier instead of λ_0 as a matter of convenience. Let $\partial L/\partial p_i = 0$, we have

$$\Delta(X)\left(-\lambda_0 - \sum_{k=1}^{K} \lambda_k g_k(s_i) - \log p_i\right) + c_i = 0, \quad i = 1, 2, \ldots, M, \tag{4.259}$$

where

$$c_i = \begin{cases} \dfrac{|s_M - s_1|}{2(M-1)} + \dfrac{|s_2 - s_1|}{2}, & i = 1 \\[3mm] \dfrac{|s_{i+1} - s_{i-1}|}{2}, & i = 2, \ldots, M-1 \\[3mm] \dfrac{|s_M - s_1|}{2(M-1)} + \dfrac{|s_M - s_{M-1}|}{2}, & i = M \end{cases} \tag{4.260}$$

Solving Eq. (4.259), we obtain the following theorem:

Theorem 8 The distribution P that maximizes the Δ-entropy subject to the constraints of Eq. (4.257) is given by

$$p_i = \exp\left(-\lambda_0 - \sum_{k=1}^{K} \lambda_k g_k(s_i) + \frac{c_i}{\Delta(X)}\right), \quad i = 1, 2, \ldots, M \tag{4.261}$$

where $\lambda_0, \lambda_1, \ldots, \lambda_K$ are determined by substituting for p_i from Eq. (4.261) into the constraints of Eq. (4.257).

For the case in which the discrete values are equally spaced, we have $c_1 = c_2 = \cdots = c_M = \Delta$, and Eq. (4.261) becomes

$$p_i = \exp\left(1 - \lambda_0 - \sum_{k=1}^{K} \lambda_k g_k(s_i)\right) \tag{4.262}$$

In this case, the maximum Δ-entropy distribution is identical to the maximum discrete entropy distribution [149].

4.7.3 Estimation of Δ-Entropy

In practical situations, the discrete values $\{s_i\}$ and probabilities $\{p_i\}$ are usually unknown, and we must estimate them from sample data $\{x_1, x_2, \ldots, x_n\}$. An immediate approach is to group the sample data into different values $\{\hat{s}_i\}$ and calculate the corresponding relative frequencies:

$$\hat{p}_i = n_i/n, \quad i = 1, 2, \ldots, M \tag{4.263}$$

where n_i denotes the number of these outcomes belonging to the value $\hat{s}_i$ with $\sum_{i=1}^{M} n_i = n$.

Based on the estimated values $\{\hat{s}_i\}$ and probabilities $\{\hat{p}_i\}$, a simple *plug-in* estimate of Δ-entropy can be obtained as

$$H_\Delta(\hat{S}, \hat{P}) = -\sum_{i=1}^{M} \hat{p}_i \log \hat{p}_i + \log\left(\sum_{i=1}^{M-1} |\hat{s}_{i+1} - \hat{s}_i| \frac{\hat{p}_i + \hat{p}_{i+1}}{2} + \frac{|\hat{s}_M - \hat{s}_1|}{M-1} \frac{\hat{p}_1 + \hat{p}_M}{2}\right) \tag{4.264}$$

where $\hat{S} = (\hat{s}_1, \hat{s}_2, \ldots, \hat{s}_M)$ and $\hat{P} = (\hat{p}_1, \hat{p}_2, \ldots, \hat{p}_M)$.

As the sample size increases, the estimated value set $\hat{S}$ will approach the true value set S with probability one, i.e., $\Pr(\hat{S} = S) = 1$, as $n \to \infty$. In fact, assuming $\{x_1, x_2, \ldots, x_n\}$ is an i.i.d. sample from the distribution P, and $p_i > 0$, $i = 1, \ldots, M$, we have

$$
\begin{aligned}
\Pr(\hat{S} \neq S) &\leq \sum_{i=1}^{M} \Pr(s_i \notin \{x_1, x_2, \ldots, x_n\}) \\
&= \sum_{i=1}^{M} \left(\prod_{j=1}^{n} \Pr(X_j \neq s_i) \right) \\
&= \sum_{i=1}^{M} (1 - p_i)^n \to 0 \quad \text{as } n \to \infty
\end{aligned}
\tag{4.265}
$$

We investigate in the following the asymptotic behavior of the Δ-entropy in random sampling. We assume for tractability that the value set S is known (or has been exactly estimated). Following a similar derivation of the asymptotic distribution for the ϕ-entropy (see Ref. [130], Chap. 2), we denote the parameter vector $\theta = (\theta_1, \theta_2, \ldots, \theta_{M-1})^T = (p_1, p_2, \ldots, p_{M-1})^T$, and rewrite Eq. (4.264) as

$$
\begin{aligned}
H_\Delta(\hat{\theta}) = &- \sum_{i=1}^{M-1} \hat{\theta}_i \log \hat{\theta}_i - \left(1 - \sum_{j=1}^{M-1} \hat{\theta}_j\right) \log\left(1 - \sum_{j=1}^{M-1} \hat{\theta}_j\right) \\
&+ \log\left(\sum_{i=1}^{M-2} |s_{i+1} - s_i| \frac{\hat{\theta}_i + \hat{\theta}_{i+1}}{2} + |s_M - s_{M-1}| \frac{\hat{\theta}_{M-1} + \left(1 - \sum_{j=1}^{M-1} \hat{\theta}_j\right)}{2} \right. \\
&\left. + \frac{|s_M - s_1|}{M-1} \frac{\hat{\theta}_1 + \left(1 - \sum_{j=1}^{M-1} \hat{\theta}_j\right)}{2} \right)
\end{aligned}
\tag{4.266}
$$

The first-order Taylor expansion of $H_\Delta(\hat{\theta})$ around θ gives

$$
H_\Delta(\hat{\theta}) = H_\Delta(\theta) + \sum_{i=1}^{M-1} \frac{\partial H_\Delta(\theta)}{\partial \theta_i} (\hat{\theta}_i - \theta_i) + o(\|\hat{\theta} - \theta\|)
\tag{4.267}
$$

where $\|\hat{\theta} - \theta\| = \sqrt{(\hat{\theta} - \theta)^T (\hat{\theta} - \theta)}$, and $\partial H_\Delta(\theta)/\partial \theta_i$ is

$$
\frac{\partial H_\Delta(\boldsymbol{\theta})}{\partial \theta_i} =
\begin{cases}
-\log\theta_i + \log\left(1 - \sum_{j=1}^{M-1}\theta_j\right) + \dfrac{s_{i+1} - s_{i-1} - (s_M - s_{M-1})}{2\Delta} - \dfrac{|s_M - s_1|}{2(M-1)\Delta}, \\[4pt]
\quad i \neq 1, M-1 \\[10pt]
-\log\theta_1 + \log\left(1 - \sum_{j=1}^{M-1}\theta_j\right) + \dfrac{s_2 - s_1 - (s_M - s_{M-1})}{2\Delta}, \quad i = 1 \\[12pt]
-\log\theta_{M-1} + \log\left(1 - \sum_{j=1}^{M-1}\theta_j\right) + \dfrac{s_{M-1} - s_{M-2}}{2\Delta} - \dfrac{|s_M - s_1|}{2(M-1)\Delta}, \\[4pt]
\quad i = M-1
\end{cases}
$$

$$(4.268)$$

where

$$
\Delta = \left(\sum_{i=1}^{M-2} |s_{i+1} - s_i| \frac{\hat{\theta}_i + \hat{\theta}_{i+1}}{2} + |s_M - s_{M-1}| \frac{\hat{\theta}_{M-1} + \left(1 - \sum_{j=1}^{M-1}\hat{\theta}_j\right)}{2} \right.
$$

$$
\left. + \frac{|s_M - s_1|}{M-1} \frac{\hat{\theta}_1 + \left(1 - \sum_{j=1}^{M-1}\hat{\theta}_j\right)}{2} \right)
$$

$$(4.269)$$

According to Ref. [130, Chap. 2], we have

$$
\sqrt{n}(\hat{\boldsymbol{\theta}} - \boldsymbol{\theta}) \xrightarrow[n\to\infty]{L} \mathcal{N}(\mathbf{0}, I_F(\boldsymbol{\theta})^{-1})
$$

$$(4.270)$$

where the inverse of the Fisher information matrix of $\boldsymbol{\theta}$ is given by $I_F(\boldsymbol{\theta})^{-1} = \text{diag}(\boldsymbol{\theta}) - \boldsymbol{\theta}\boldsymbol{\theta}^T$. Then $\sqrt{n}\|\hat{\boldsymbol{\theta}} - \boldsymbol{\theta}\|$ is bounded in probability, and

$$
\sqrt{n}(o(\|\hat{\boldsymbol{\theta}} - \boldsymbol{\theta}\|)) \xrightarrow[n\to\infty]{P} 0
$$

$$(4.271)$$

And hence, random variables $\sqrt{n}(H_\Delta(\hat{\boldsymbol{\theta}}) - H_\Delta(\boldsymbol{\theta}))$ and $\sqrt{n}\sum_{i=1}^{M-1} \frac{\partial H_\Delta(\boldsymbol{\theta})}{\partial \theta_i}(\hat{\theta}_i - \theta_i)$ have the same asymptotic distribution, and we have the following theorem.

Theorem 9 The estimate $H_\Delta(S,\hat{\boldsymbol{P}})$, obtained by replacing the $\{p_i\}$ by their relative frequencies $\{\hat{p}_i\}$, in a random sample of size n, satisfies

$$
\sqrt{n}(H_\Delta(S,\hat{\boldsymbol{P}}) - H_\Delta(S,P)) \xrightarrow[n\to\infty]{L} \mathcal{N}(0, \boldsymbol{U}^T I_F(\boldsymbol{\theta})^{-1}\boldsymbol{U})
$$

$$(4.272)$$

provided $U^T I_F(\theta)^{-1} U > 0$, where $\theta = (p_1, p_2, \ldots, p_{M-1})^T$, $I_F(\theta)^{-1} = \text{diag}(\theta) - \theta\theta^T$, and

$$U = (\partial H_\Delta(\theta)/\partial\theta_1, \partial H_\Delta(\theta)/\partial\theta_2, \ldots, \partial H_\Delta(\theta)/\partial\theta_{M-1})^T \tag{4.273}$$

where $\partial H_\Delta(\theta)/\partial\theta_i$ is calculated as (4.268).

The afore-discussed plug-in estimator of the Δ-entropy has close relationships with certain estimators of the differential entropy.

4.7.3.1   Relation to KDE-based Differential Entropy Estimator

Suppose $\{x_1, x_2, \ldots, x_n\}$ are samples from a discrete random variable X. We rewrite the plug-in estimate (4.264) as

$$H_\Delta(\hat{S}, \hat{P}) = -\sum_{i=1}^{M} \hat{p}_i \log \hat{p}_i + \log \hat{\Delta} \tag{4.274}$$

where $\hat{\Delta} = \sum_{i=1}^{M-1} |\hat{s}_{i+1} - \hat{s}_i|((\hat{p}_i + \hat{p}_{i+1})/2) + (|\hat{s}_M - \hat{s}_1|/(M-1))((\hat{p}_1 + \hat{p}_M)/2)$. Denote $\hat{\Delta}_{\min} = \min_{i=1,\cdots,M-1} |\hat{s}_{i+1} - \hat{s}_i|$, and let $\tau = \hat{\Delta}/\hat{\Delta}_{\min}$, we construct another set of samples:

$$\{x'_1, x'_2, \ldots, x'_n\} = \{\tau x_1, \tau x_2, \ldots, \tau x_n\} \tag{4.275}$$

which are samples from the discrete random variable τX, and satisfy

$$\forall x'_i \neq x'_j, |x'_i - x'_j| \geq \hat{\Delta} \tag{4.276}$$

Now we consider $\{x'_1, x'_2, \ldots, x'_n\}$ as samples from a "continuous" random variable X'. The PDF of X' can be estimated by the KDE approach:

$$\hat{p}(x') = \frac{1}{n}\sum_{i=1}^{n} K(x' - x'_i) \tag{4.277}$$

The kernel function satisfies $K \geq 0$ and $\int_{-\infty}^{\infty} K(x)dx = 1$. If the kernel function is selected as the following uniform kernel:

$$K_{\hat{\Delta}}(x) = \begin{cases} 1/\hat{\Delta}, & x \in [-\hat{\Delta}/2, \hat{\Delta}/2] \\ 0 & \text{otherwise} \end{cases} \tag{4.278}$$

then Eq. (4.277) becomes

$$\hat{p}(x') = \frac{1}{n}\sum_{j=1}^{n}K_{\hat{\Delta}}(x' - x'_j)$$

$$= \frac{1}{n}\sum_{i=1}^{M}n_i K_{\hat{\Delta}}(x' - s'_i) \tag{4.279}$$

$$\overset{(a)}{=}\begin{cases}\dfrac{p_i}{\hat{\Delta}}, & x' \in \left[s'_i - \hat{\Delta}/2, s'_i + \hat{\Delta}/2\right] \\[2mm] 0 & \text{otherwise}\end{cases}$$

where (a) follows from $\hat{p}_i = n_i/n$, and $\forall\, x'_i \neq x'_j,\ |x'_i - x'_j| \geq \hat{\Delta}$. The differential entropy of X' can then be estimated as

$$\begin{aligned}\hat{h}(X') &= -\int_{-\infty}^{\infty}\hat{p}(x')\log\hat{p}(x')\mathrm{d}x' \\ &= -\sum_{i=1}^{M}\int_{s'_i-\hat{\Delta}/2}^{s'_i+\hat{\Delta}/2}\hat{p}(x')\log\hat{p}(x')\mathrm{d}x' \\ &= -\sum_{i=1}^{M}\int_{s'_i-\hat{\Delta}/2}^{s'_i+\hat{\Delta}/2}\frac{p_i}{\hat{\Delta}}\log\frac{p_i}{\hat{\Delta}}\mathrm{d}x' \\ &= -\sum_{i=1}^{M}p_i\log p_i + \log\hat{\Delta} = H_{\Delta}(\hat{S},\hat{P})\end{aligned} \tag{4.280}$$

As a result, the plug-in estimate of the Δ-entropy is identical to a uniform kernel-based estimate of the differential entropy from the scaled samples (4.275).

4.7.3.2 Relation to Sample-Spacing Based Differential Entropy Estimator

The plug-in estimate of the Δ-entropy also has a close connection with the sample-spacing based estimate of the differential entropy. Suppose the sample data are different from each other, and have been rearranged in an increasing order: $x_1 < x_2 < \cdots < x_n$, the m-spacing estimate is given by [214]

$$\hat{h}_m(X) = \frac{1}{n}\sum_{i=1}^{n-m}\log\left(\frac{n}{m}(x_{i+m} - x_i)\right) \tag{4.281}$$

where $m \in \mathbb{N}$ and $m < n$. If $m = 1$, we obtain the one-spacing estimate:

$$\hat{h}_1(X) = \frac{1}{n}\sum_{i=1}^{n-1}\log(n(x_{i+1} - x_i)) \tag{4.282}$$

On the other hand, based on the samples one can estimate the value set and corresponding probabilities:

$$\begin{cases} \hat{S} = (x_1, x_2, \ldots, x_n) \\ \hat{P} = (1/n, 1/n, \ldots, 1/n) \end{cases} \tag{4.283}$$

Then the plug-in estimate of the Δ-entropy will be

$$H_\Delta(\hat{S},\hat{P}) = -\sum_{i=1}^{n} \frac{1}{n} \log \frac{1}{n} + \log\left(\sum_{i=1}^{n-1}(x_{i+1} - x_i)\frac{1}{n} + \frac{(x_n - x_1)}{n-1}\frac{1}{n}\right) \tag{4.284}$$

It follows that

$$\begin{aligned} H_\Delta(\hat{S},\hat{P}) &= -\sum_{i=1}^{n} \frac{1}{n} \log \frac{1}{n} + \log\left(\sum_{i=1}^{n-1}(x_{i+1} - x_i)\frac{1}{n} + \frac{(x_n - x_1)}{n-1}\frac{1}{n}\right) \\ &\overset{(b)}{\geq} \frac{1}{n}\sum_{i=1}^{n} \log n + \frac{1}{n}\left(\sum_{i=1}^{n-1}\log(x_{i+1} - x_i) + \log\frac{(x_n - x_1)}{n-1}\right) \\ &= \frac{1}{n}\sum_{i=1}^{n-1}\log(n(x_{i+1} - x_i)) + \frac{1}{n}\log\frac{n(x_n - x_1)}{n-1} \\ &= \hat{h}_1(X) + \frac{1}{n}\log\frac{n(x_n - x_1)}{n-1} \end{aligned} \tag{4.285}$$

where (b) comes from the concavity of the logarithm function. If $\{x_i\}$ is bounded, we have

$$\lim_{n\to\infty} H_\Delta(\hat{S},\hat{P}) \geq \lim_{n\to\infty}\left(\hat{h}_1(X) + \frac{1}{n}\log\frac{n(x_n - x_1)}{n-1}\right) = \lim_{n\to\infty} \hat{h}_1(X) \tag{4.286}$$

In this case, the plug-in estimate of Δ-entropy provides an asymptotic upper bound on the one-spacing entropy estimate.

4.7.4 Application to System Identification

The Δ-entropy can be used as an optimality criterion in system identification, especially when error signal is distributed on a countable value set (which is usually unknown and varying with time) [15]. A typical example is the system identification with quantized input/output (I/O) data, as shown in Figure 4.32, where $\bar{x}_k$ and $\bar{z}_k$

---

[15] For the case in which the error distribution is continuous, one can still use the Δ-entropy as the optimization criterion if classifying the errors into groups and obtaining the quantized data.

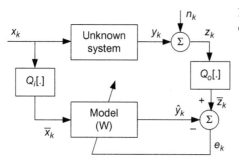

Figure 4.32 System identification with quantized I/O data.

represent the quantized I/O observations, obtained via I/O quantizers Q_i and Q_o. With uniform quantization, $\bar{x}_k$ and $\bar{z}_k$ can be expressed as

$$
\begin{cases}
\bar{x}_k = \lceil x_k/q_i + 1/2 \rceil \times q_i \\
\bar{z}_k = \lceil z_k/q_o + 1/2 \rceil \times q_o
\end{cases}
\tag{4.287}
$$

where q_i and q_o denote the quantization box-sizes, $\lceil x \rceil$ gives the largest integer that is less than or equal to x.

In practice, Δ-entropy cannot be, in general, analytically computed, since the error's values and corresponding probabilities are unknown. In this case, we need to estimate the Δ-entropy by the plug-in method as discussed previously. Traditional gradient-based methods, however, cannot be used to solve the Δ-entropy minimization problem, since the objective function is usually not differentiable. Thus, we have to resort to other methods, such as the estimation of distribution algorithms (EDAs) [215], a new class of evolutionary algorithms (EAs), although they are usually more computationally complex. The EDAs use the probability model built from the objective function to generate the promising search points instead of crossover and mutation as done in traditional GAs. Some theoretical results related to the convergence and time complexity of EDAs can be found in Ref. [215]. Table 4.8 presents the EDA-based identification algorithm with Δ-entropy criterion.

Usually, we use a Gaussian model with diagonal covariance matrix (GM/DCM) [215] to estimate the density function $f_g(W)$ of the gth generation. With GM/DCM model, we have

$$
f_g(W) = \prod_{j=1}^{m} \frac{1}{\sqrt{2\pi}\sigma_j^{(g)}} \exp\left(-(w_j - \mu_j^{(g)})^2/(2(\sigma_j^{(g)})^2)\right)
\tag{4.288}
$$

where the means $\mu_j^{(g)}$ and the deviations $\sigma_j^{(g)}$ can be estimated as

$$
\begin{cases}
\mu_j^{(g)} = \frac{1}{N}\sum_{l=1}^{N} W_{B(l)}^{(g-1)}(j) \\[4mm]
\sigma_j^{(g)} = \sqrt{\frac{1}{N}\sum_{l=1}^{N}(W_{B(l)}^{(g-1)}(j) - \mu_j^{(g)})^2}
\end{cases}
\tag{4.289}
$$

Table 4.8 EDA Based Identification Algorithm with Δ-entropy Criterion

1. **BEGIN**
2. Generate R individuals $A_0 = \{W_1^{(0)}, W_2^{(0)}, \ldots, W_R^{(0)}\}$ randomly from parameter space, $g \leftarrow 0$
3. **WHILE** the final stopping criterion is not met **DO**
4. $g \leftarrow g + 1$
5. For each parameter vector in A_{g-1}, estimate the error's Δ-entropy using a training data set
6. Select $N(N \leq R)$ promising individuals $B_g = \{W_{B(1)}^{(g-1)}, W_{B(2)}^{(g-1)}, \ldots, W_{B(N)}^{(g-1)}\}$ from A_{g-1} according to the truncation selection method (using Δ-entropy as the fitness function)[16]
7. Estimate the PDF $f_g(W)$ based on the statistical information extracted from the selected N individuals B_g
8. Sample R individuals $A_g = \{W_1^{(g)}, W_2^{(g)}, \ldots, W_R^{(g)}\}$ from $f_g(W)$
9. **END WHILE**
10. Calculate the estimated parameter: $W(g) = (1/N) \sum_{n=1}^{N} W_{B(n)}^{(g-1)}$
11. **END**

In the following, two simple examples are presented to demonstrate the performance of the above algorithm. In all of the simulations below, we set $R = 100$ and $N = 30$.

First, we consider the system identification based on quantized I/O data, where the unknown system and the parametric model are both two-tap FIR filters, i.e.,

$$\begin{cases} z_k = w_1^* x_k + w_2^* x_{k-1} \\ y_k = w_1 \bar{x}_k + w_2 \bar{x}_{k-1} \end{cases} \tag{4.290}$$

The true weight vector of the unknown system is $W^* = [1.0, 0.5]^T$, and the initial weight vector of the model is $W_0 = [0, 0]^T$. The input signal and the additive noise are both white Gaussian processes with variances 1.0 and 0.04, respectively. The number of the training data is 500. In addition, the quantization box-size q_i and q_o are equal. We compare the performance among three entropy criteria: Δ-entropy, differential entropy[17] and discrete entropy. For different quantization sizes, the average evolution curves of the weight error norm over 100 Monte Carlo runs are shown in Figure 4.33. We can see the Δ-entropy criterion achieves the best performance, and the discrete entropy criterion fails to converge (discrete entropy cannot constrain the dispersion of the error value). When the quantization size becomes smaller, the performance of the differential entropy approaches that of the Δ-entropy. This agrees with the limiting relationship between the Δ-entropy and the differential entropy.

The second example illustrates that the Δ-entropy criterion may yield approximately an unbiased solution even if the input and output data are both corrupted by

---

[16] The truncation selection is a widely used selection method in EDAs. In the truncation selection, individuals are sorted according to their objective function (or fitness function) values and only the best individuals are selected.

[17] Strictly speaking, the differential entropy criterion is invalid in this example, because the error is discrete-valued. However, in the simulation one can still adopt the empirical differential entropy.

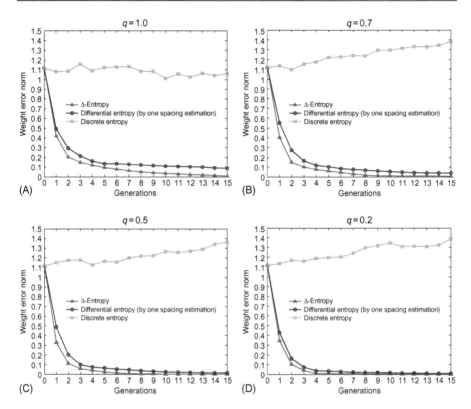

Figure 4.33 Evolution curves of the weight error norm for different entropy criteria (adopted from Ref. [211]).

noises. Consider again the identification of a two-tap FIR filter in which $G^*(z) = w_1^* + w_2^* z^{-1} = 1.0 + 0.5z^{-1}$. We assume the input signal x_k, input noise $n_{1,k}$, and the output noise $n_{2,k}$ are all zero-mean white Bernoulli processes with distributions below

$$\begin{cases} \Pr\{x_k = \sigma_x\} = 0.5, & \Pr\{x_k = -\sigma_x\} = 0.5 \\ \Pr\{n_{1,k} = \sigma_{n_1}\} = 0.5, & \Pr\{n_{1,k} = -\sigma_{n_1}\} = 0.5 \\ \Pr\{n_{2,k} = \sigma_{n_2}\} = 0.5, & \Pr\{n_{2,k} = -\sigma_{n_2}\} = 0.5 \end{cases} \tag{4.291}$$

where σ_x, σ_{n_1}, and σ_{n_2} denote, respectively, the standard deviations of x_k, $n_{1,k}$, and $n_{2,k}$. In the simulation we set $\sigma_x = 1.0$, and the number of training data is 500. Simulation results over 100 Monte Carlo runs are listed in Tables 4.9 and 4.10. For comparison purpose, we also present the results obtained using MSE criterion. As one can see, the Δ-entropy criterion produces nearly unbiased estimates under various SNR conditions, whereas the MSE criterion yields biased solution especially when the input noise power increasing.

Table 4.9 Simulation Results for Different σ_{n_1} ($\sigma_{n_2} = 0.1$)

σ_{n_1}	Δ-entropy		MSE	
	w_1	w_2	w_1	w_2
0.1	1.0000 ± 0.0008	0.4999 ± 0.0007	0.9896 ± 0.0071	0.4952 ± 0.0067
0.2	0.9997 ± 0.0015	0.4995 ± 0.0015	0.9609 ± 0.0098	0.4800 ± 0.0097
0.3	0.9994 ± 0.0016	0.4993 ± 0.0017	0.9192 ± 0.0118	0.4617 ± 0.0140
0.4	0.9991 ± 0.0019	0.4991 ± 0.0019	0.8629 ± 0.0129	0.4316 ± 0.0163
0.5	0.9980 ± 0.0039	0.4972 ± 0.0077	0.8015 ± 0.0146	0.4016 ± 0.0200

(adopted from Ref. [211])

Table 4.10 Simulation Results for Different σ_{n_2} ($\sigma_{n_1} = 0.1$)

σ_{n_2}	Δ-entropy		MSE	
	w_1	w_2	w_1	w_2
0.1	1.0000 ± 0.0008	0.4999 ± 0.0007	0.9896 ± 0.0071	0.4952 ± 0.0067
0.2	0.9999 ± 0.0009	0.4999 ± 0.0008	0.9887 ± 0.0096	0.4949 ± 0.0089
0.3	0.9999 ± 0.0012	0.4998 ± 0.0011	0.9908 ± 0.0139	0.4943 ± 0.0142
0.4	0.9999 ± 0.0020	0.4998 ± 0.0017	0.9880 ± 0.0192	0.4948 ± 0.0208
0.5	1.0001 ± 0.0039	0.4998 ± 0.0024	0.9926 ± 0.0231	0.4946 ± 0.0235

(adopted from Ref. [211])

4.8   System Identification with MCC

Correntropy is closely related to Renyi's quadratic entropy. With Gaussian kernel, correntropy is a localized similarity measure between two random variables: when two points are close, the correntropy induced metric (CIM) behaves like an L2 norm; outside of the L2 zone CIM behaves like an L1 norm; as two points are further apart, the metric approaches L0 norm [137]. This property makes the MCC a robust adaptation criterion in presence of non-Gaussian impulsive noise. At the end of this chapter, we briefly discuss the application of the MCC criterion to system identification.

Consider a general scheme of system identification as shown in Figure 4.1. The objective of the identification is to optimize a criterion function (or cost function) in such a way that the model output $\hat{y}_k$ resembles as closely as possible the measured output z_k. Under the MCC criterion, the cost function that we want to maximize is the correntropy between the measured output and the model output, i.e.,

$$J = E[\kappa_\sigma(z_k, \hat{y}_k)]$$

$$= \frac{1}{\sqrt{2\pi}\sigma} E\left[\exp\left(-\frac{(z_k - \hat{y}_k)^2}{2\sigma^2}\right)\right]$$

$$= \frac{1}{\sqrt{2\pi}\sigma} E\left[\exp\left(-\frac{e_k^2}{2\sigma^2}\right)\right] \tag{4.292}$$

In practical applications, one often uses the following empirical correntropy as the cost function:

$$\hat{J} = \frac{1}{\sqrt{2\pi}\sigma} \cdot \frac{1}{L} \sum_{i=k-L+1}^{k} \exp\left(-\frac{e_i^2}{2\sigma^2}\right) \tag{4.293}$$

Then a gradient-based identification algorithm can be easily derived as follows:

$$W_k = W_{k-1} + \eta \sum_{i=k-L+1}^{k} \exp\left(-\frac{e_i^2}{2\sigma^2}\right) e_i \frac{\partial \hat{y}_i}{\partial W} \tag{4.294}$$

When $L = 1$, the above algorithm becomes a stochastic gradient-based (LMS-like) algorithm:

$$W_k = W_{k-1} + \eta \exp\left(-\frac{e_k^2}{2\sigma^2}\right) e_k \frac{\partial \hat{y}_k}{\partial W} \tag{4.295}$$

In the following, we present two simulation examples of FIR identification to demonstrate the performance of MCC criterion, and compare it with the performance of MSE and MEE.

In the first example, the weight vector of the plant is [138]

$$W^* = [0.1, 0.2, 0.3, 0.4, 0.5, 0.4, 0.3, 0.2, 0.1]^T \tag{4.296}$$

The input signal is a white Gaussian process with zero mean and unit variance. The noise distribution is a mixture of Gaussian:

$$0.95\mathcal{N}(0, 10^{-4}) + 0.05\mathcal{N}(0, 10) \tag{4.297}$$

In this distribution, the Gaussian density with variance 10 creates strong outliers. The kernel sizes for the MCC and the MEE are set at 2.0. The step-sizes for the three identification criteria are chosen such that when the observation noise is Gaussian, their performance is similar in terms of the weight SNR (WSNR),

$$\text{WSNR} = 10 \log_{10}\left(\frac{W^{*T} W^*}{(W^* - W_k)^T (W^* - W_k)}\right) \tag{4.298}$$

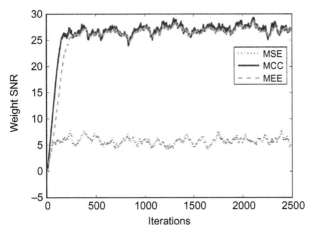

Figure 4.34 WSNR of three identification criteria in impulsive measurement noise (adopted from Ref. [138]).

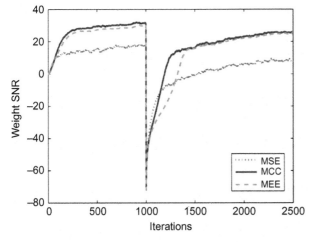

Figure 4.35 WSNR of three identification criteria while tracking a time-varying system in impulsive measurement noise (adopted from Ref. [138]).

Figure 4.34 shows the performance in the presence of impulsive noise. One can see the MCC criterion achieves a more robust performance.

In the second example, the plant has a time-varying transfer function, where the weight vector is changing as follows [138]:

$$W_k^* = 2\left(1 + \frac{k}{1000}\right)u(1000 - k)W^* + \left(-1 + \frac{k}{1000}\right)u(k - 1000)W^* \quad (4.299)$$

where $u(.)$ is the unit step function. The simulation results are shown in Figure 4.35. Once again the MCC criterion performs better in impulsive noise environment.

Appendix H: Vector Gradient and Matrix Gradient

In many of system identification applications, we often encounter a cost function that we want to maximize (or minimize) with respect to a vector or matrix. To accomplish this optimization, we usually need to find a vector or matrix derivative.

Derivative of Scalar with Respect to Vector

If J is a scalar, $\boldsymbol{\theta} = \begin{bmatrix} \theta_1 & \theta_2 & \cdots & \theta_m \end{bmatrix}^T$ is an $m \times 1$ vector, then

$$\frac{\partial J}{\partial \boldsymbol{\theta}} \triangleq \begin{bmatrix} \frac{\partial J}{\partial \theta_1}, & \frac{\partial J}{\partial \theta_2}, & \cdots & \frac{\partial J}{\partial \theta_m} \end{bmatrix}^T \tag{H.1}$$

$$\frac{\partial J}{\partial \boldsymbol{\theta}^T} \triangleq \begin{bmatrix} \frac{\partial J}{\partial \theta_1}, & \frac{\partial J}{\partial \theta_2}, & \cdots & \frac{\partial J}{\partial \theta_m} \end{bmatrix} \tag{H.2}$$

Derivative of Scalar with Respect to Matrix

If J is a scalar, $\boldsymbol{M}$ is an $m \times n$ matrix, $\boldsymbol{M} = [m_{ij}]$, then

$$\frac{\partial J}{\partial \mathbf{M}} \triangleq \begin{bmatrix} \dfrac{\partial J}{\partial m_{11}} & \dfrac{\partial J}{\partial m_{12}} & \cdots & \dfrac{\partial J}{\partial m_{1n}} \\ \dfrac{\partial J}{\partial m_{21}} & \dfrac{\partial J}{\partial m_{22}} & \cdots & \dfrac{\partial J}{\partial m_{2n}} \\ \vdots & \vdots & \cdots & \vdots \\ \dfrac{\partial J}{\partial m_{m1}} & \dfrac{\partial J}{\partial m_{m2}} & \cdots & \dfrac{\partial J}{\partial m_{mn}} \end{bmatrix} \tag{H.3}$$

Derivative of Vector with Respect to Vector

If $\boldsymbol{\alpha}$ and $\boldsymbol{\theta}$ are, respectively, $n \times 1$ and $m \times 1$ vectors, then

$$\frac{\partial \boldsymbol{\alpha}^T}{\partial \boldsymbol{\theta}} \triangleq \begin{bmatrix} \dfrac{\partial \alpha_1}{\partial \boldsymbol{\theta}}, & \dfrac{\partial \alpha_2}{\partial \boldsymbol{\theta}}, & \cdots, & \dfrac{\partial \alpha_n}{\partial \boldsymbol{\theta}} \end{bmatrix} = \begin{bmatrix} \dfrac{\partial \alpha_1}{\partial \theta_1} & \dfrac{\partial \alpha_2}{\partial \theta_1} & \cdots & \dfrac{\partial \alpha_n}{\partial \theta_1} \\ \dfrac{\partial \alpha_1}{\partial \theta_2} & \dfrac{\partial \alpha_2}{\partial \theta_2} & \cdots & \dfrac{\partial \alpha_n}{\partial \theta_2} \\ \vdots & \vdots & \cdots & \vdots \\ \dfrac{\partial \alpha_1}{\partial \theta_m} & \dfrac{\partial \alpha_2}{\partial \theta_m} & \cdots & \dfrac{\partial \alpha_n}{\partial \theta_m} \end{bmatrix} \tag{H.4}$$

$$\frac{\partial \boldsymbol{\alpha}}{\partial \boldsymbol{\theta}^T} \triangleq \left[\frac{\partial \alpha_1}{\partial \boldsymbol{\theta}}, \quad \frac{\partial \alpha_2}{\partial \boldsymbol{\theta}}, \quad \cdots, \quad \frac{\partial \alpha_n}{\partial \boldsymbol{\theta}}\right]^T = \begin{bmatrix} \dfrac{\partial \alpha_1}{\partial \theta_1} & \dfrac{\partial \alpha_1}{\partial \theta_2} & \cdots & \dfrac{\partial \alpha_1}{\partial \theta_m} \\[2mm] \dfrac{\partial \alpha_2}{\partial \theta_1} & \dfrac{\partial \alpha_2}{\partial \theta_2} & \cdots & \dfrac{\partial \alpha_2}{\partial \theta_m} \\[1mm] \vdots & \vdots & \cdots & \vdots \\[1mm] \dfrac{\partial \alpha_n}{\partial \theta_1} & \dfrac{\partial \alpha_n}{\partial \theta_2} & \cdots & \dfrac{\partial \alpha_n}{\partial \theta_m} \end{bmatrix} \tag{H.5}$$

Second Derivative (Hessian matrix) of Scalar with Respect to Vector

$$\frac{\partial^2 J}{\partial \boldsymbol{\theta}^2} \triangleq \frac{\partial}{\partial \boldsymbol{\theta}^T}\left(\frac{\partial J}{\partial \boldsymbol{\theta}}\right) = \frac{\partial}{\partial \boldsymbol{\theta}}\left(\frac{\partial J}{\partial \boldsymbol{\theta}^T}\right) = \begin{bmatrix} \dfrac{\partial^2 J}{\partial^2 \theta_1} & \dfrac{\partial^2 J}{\partial \theta_1 \partial \theta_2} & \cdots & \dfrac{\partial^2 J}{\partial \theta_1 \partial \theta_m} \\[2mm] \dfrac{\partial^2 J}{\partial \theta_2 \partial \theta_1} & \dfrac{\partial^2 J}{\partial^2 \theta_2} & \cdots & \dfrac{\partial^2 J}{\partial \theta_2 \partial \theta_m} \\[1mm] \vdots & \vdots & \cdots & \vdots \\[1mm] \dfrac{\partial^2 J}{\partial \theta_m \partial \theta_1} & \dfrac{\partial^2 J}{\partial \theta_m \partial \theta_2} & \cdots & \dfrac{\partial^2 J}{\partial^2 \theta_m} \end{bmatrix} \tag{H.6}$$

With the above definitions, there are some basic results:

1. $\frac{\partial}{\partial \boldsymbol{\theta}}[c_1 J_1 + c_2 J_2] = c_1 \frac{\partial J_1}{\partial \boldsymbol{\theta}} + c_2 \frac{\partial J_2}{\partial \boldsymbol{\theta}}$, where J_1 and J_2 are scalars, $c_1, c_2 \in \mathbb{R}$ are constants.

2. $\frac{\partial}{\partial \boldsymbol{\theta}}[J_1 J_2] = \frac{\partial J_1}{\partial \boldsymbol{\theta}} J_2 + J_1 \frac{\partial J_2}{\partial \boldsymbol{\theta}}$.

3. $\frac{\partial}{\partial \boldsymbol{\theta}}\left[\frac{J_1}{J_2}\right] = \frac{1}{J_2^2}\left[J_2 \frac{\partial J_1}{\partial \boldsymbol{\theta}} - J_1 \frac{\partial J_2}{\partial \boldsymbol{\theta}}\right]$.

4. $\frac{\partial}{\partial \boldsymbol{\theta}}\left[\boldsymbol{\alpha}^T \boldsymbol{\theta}\right] = \frac{\partial}{\partial \boldsymbol{\theta}}\left[\boldsymbol{\theta}^T \boldsymbol{\alpha}\right] = \boldsymbol{\alpha}$, where $\boldsymbol{\alpha}$ and $\boldsymbol{\theta}$ are both $m \times 1$ vectors, and $\boldsymbol{\alpha}$ and $\boldsymbol{\theta}$ are independent.

5. $\frac{\partial}{\partial \boldsymbol{\theta}}\left[\boldsymbol{\theta}^T A\right] = A$, $\frac{\partial}{\partial \boldsymbol{\theta}}\left[\boldsymbol{\theta}^T A \boldsymbol{\theta}\right] = A\boldsymbol{\theta} + A^T \boldsymbol{\theta}$, where A is a matrix independent of $\boldsymbol{\theta}$.

6. Let $\boldsymbol{\alpha}$ and $\boldsymbol{\theta}$ be respectively $n \times 1$ and $m \times 1$ vectors, A and B be respectively $n \times m$ and $n \times n$ constant matrices. Then

$$\frac{\partial}{\partial \boldsymbol{\theta}}\left\{(\boldsymbol{\alpha} - A\boldsymbol{\theta})^T B(\boldsymbol{\alpha} - A\boldsymbol{\theta})\right\} = -2A^T B(\boldsymbol{\alpha} - A\boldsymbol{\theta})$$

7. If A is a $m \times m$ matrix with independent elements, then

$$\frac{\partial \det[A]}{\partial A} = \det[A][A^{-1}]^T$$

5 System Identification Under Information Divergence Criteria

The fundamental contribution of information theory is to provide a unified framework for dealing with the notion of information in a precise and technical sense. Information, in a technical sense, can be quantified in a unified manner by using the Kullback−Leibler information divergence (KLID). Two information measures, Shannon's entropy and mutual information are special cases of KL divergence [43]. The use of probability in system identification is also shown to be equivalent to measuring KL divergence between the actual and model distributions. In parameter estimation, the KL divergence for inference is consistent with common statistical approaches, such as the maximum likelihood (ML) estimation. Based on the KL divergence, Akaike derived the well-known Akaike's information criterion (AIC), which is widely used in the area of model selection. Another important model selection criterion, the minimum description length, first proposed by Rissanen in 1978, is also closely related to the KL divergence. In identification of stationary Gaussian processes, it has been shown that the optimal solution to an approximation problem for Gaussian random variables with the divergence criterion is identical to the main step of the subspace algorithm [123].

There are many definitions of information divergence, but in this chapter our focus is mainly on the KLID. In most cases, the extension to other definitions is straightforward.

5.1 Parameter Identifiability Under KLID Criterion

The identifiability arises in the context of system identification, indicating whether or not the unknown parameter can be uniquely identified from the observation of the system. One would not select a model structure whose parameters cannot be identified, so the problem of identifiability is crucial in the procedures of system identification. There are many concepts of identifiability. Typical examples include Fisher information−based identifiability [216], least squares (LS) identifiability [217], consistency-in-probability identifiability [218], transfer function−based identifiability [219], and spectral density−based identifiability [219]. In the following, we discuss the fundamental problem of system parameter identifiability under KLID criterion.

System Parameter Identification. DOI: http://dx.doi.org/10.1016/B978-0-12-404574-3.00005-1
© 2013 Tsinghua University Press Ltd. Published by Elsevier Inc. All rights reserved.

5.1.1   Definitions and Assumptions

Let $\{y_k\}_{k=1}^{\infty}$ $(y_k \in \mathbb{R}^m)$ be a sequence of observations with joint probability density functions (PDFs) $f_\theta(y^n)$, $n = 1, 2, \ldots$, where $y^n = (y_1^T, \ldots, y_n^T)^T$ is a mn-dimensional column vector, $\theta \in \Theta$ is a d-dimensional parameter vector, and $\Theta \subset \mathbb{R}^d$ is the parameter space. Let θ_0 be the true parameter. The KLID between $f_{\theta_0}(y^n)$ and $f_\theta(y^n)$ will be

$$
\begin{aligned}
D_{KL}^n(\theta_0 \| \theta) &= D_{KL}(f_{\theta_0}(y^n) \| f_\theta(y^n)) \\
&= \int f_{\theta_0}(y^n) \log \frac{f_{\theta_0}(y^n)}{f_\theta(y^n)} dy^n \\
&= E_{\theta_0} \left(\log \frac{f_{\theta_0}(y^n)}{f_\theta(y^n)} \right)
\end{aligned}
\tag{5.1}
$$

where E_{θ_0} denotes the expectation of the bracketed quantity taken with respect to the actual parameter value θ_0. Based on the KLID, a natural way of parameter identification is to look for a parameter $\theta \in \Theta$, such that the KLID of Eq. (5.1) is minimized, that is,

$$
\theta_{KL}^{(n)} = \arg \min_{\theta \in \Theta} D_{KL}^n(\theta_0 \| \theta)
\tag{5.2}
$$

An important question that arises in the context of such identification problem is whether or not the parameter θ can be uniquely determined. This is the parameter identifiability problem. Assume θ_0 lies in Θ (hence $\min_\theta D_{KL}^n(\theta_0 \| \theta) = 0$). The notion of identifiability under KLID criterion can then be defined as follows.

Definition 5.1 The parameter set Θ is said to be KLID-identifiable at $\theta \in \Theta$, if and only if $\exists M \in \mathbb{N}$, $\forall \alpha \in \Theta$, $D_{KL}^M(\theta \| \alpha) = 0$ implies $\alpha = \theta$.

By the definition, if parameter set Θ is KLID-identifiable at θ (we also say θ is KLID-identifiable), then for any $\alpha \in \Theta$, $\alpha \neq \theta$, we have $D_{KL}^M(\theta \| \alpha) \neq 0$, and hence $f_\alpha(y^M) \neq f_\theta(y^M)$. Therefore, any change in the parameter yields changes in the output density.

The identifiability can also be defined in terms of the information divergence rate.

Definition 5.2 The parameter set Θ is said to be KLIDR-identifiable at $\theta \in \Theta$, if and only if $\forall \alpha \in \Theta$, the KL information divergence rate (KLIDR) $\overline{D}_{KL}(\theta \| \alpha) = \lim_{n \to \infty} \frac{1}{n} D_{KL}^n(\theta \| \alpha)$ exists, and $\overline{D}_{KL}(\theta \| \alpha) = 0$ implies $\alpha = \theta$.

Let $B(\theta, \varepsilon) \triangleq \{x | x \in \Theta, \|x - \theta\| < \varepsilon\}$ be the ε-neighborhood of θ, where $\|\cdot\|$ denotes the Euclidean norm. The local KLID (or local KLIDR)-identifiability is defined as follows.

Definition 5.3 The parameter set Θ is said to be locally KLID (or locally KLIDR)-identifiable at $\theta \in \Theta$, if and only if there exists $\varepsilon > 0$, such that $\forall \alpha \in B(\theta, \varepsilon)$, $D_{KL}^M(\theta \| \alpha) = 0$ (or $\overline{D}_{KL}(\theta \| \alpha) = 0$) implies $\alpha = \theta$.

Here, we give some assumptions that will be used later on.

Assumption 5.1 $\forall M \in \mathbb{N}$, $\forall \theta, \alpha \in \Theta$, the KLID $D_{KL}^M(\theta \| \alpha)$ always exists.

Remark: Let $(\Omega, \mathscr{B}(\Omega), P_\theta)$ be the probability space of the output sequence y^M with parameter $\theta \in \Theta$, where $(\Omega, \mathscr{B}(\Omega))$ is the related measurable space, and $P_\theta : \mathscr{B}(\Omega) \to \mathbb{R}_+$ is the probability measure. P_θ is said to be absolutely continuous with respect to P_α, denoted by $P_\theta \prec P_\alpha$, if $P_\theta(A) = 0$ for every $A \in \mathscr{B}(\Omega)$ such that $P_\alpha(A) = 0$. Clearly, the existence of $D_{KL}^M(\theta \| \alpha)$ implies $P_\theta \prec P_\alpha$. Thus by Assumption 5.1, $\forall \theta, \alpha \in \Theta$, we have $P_\theta \prec P_\alpha$, $P_\alpha \prec P_\theta$.

Assumption 5.2 The density function $f_\theta(y^M)$ is at least twice continuously differentiable with respect to $\theta \in \Theta$, and $\forall \theta, \alpha \in \Theta$, the following interchanges between integral (or limitation) and derivative are permissible:

$$
\begin{cases}
\dfrac{\partial}{\partial \alpha} \displaystyle\int f_\theta(y^M) \log f_\alpha(y^M) dy^M = \int f_\theta(y^M) \dfrac{\partial}{\partial \alpha}\{\log f_\alpha(y^M)\} dy^M \\[4mm]
\dfrac{\partial^2}{\partial \alpha^2} \displaystyle\int f_\theta(y^M) \log f_\alpha(y^M) dy^M = \int f_\theta(y^M) \dfrac{\partial^2}{\partial \alpha^2}\{\log f_\alpha(y^M)\} dy^M
\end{cases}
\tag{5.3}
$$

$$
\begin{cases}
\dfrac{\partial}{\partial \alpha} \lim_{n \to \infty} \dfrac{1}{n} D_{KL}^n(\theta \| \alpha) = \lim_{n \to \infty} \dfrac{1}{n} \dfrac{\partial}{\partial \alpha} D_{KL}^n(\theta \| \alpha) \\[4mm]
\dfrac{\partial^2}{\partial \alpha^2} \lim_{n \to \infty} \dfrac{1}{n} D_{KL}^n(\theta \| \alpha) = \lim_{n \to \infty} \dfrac{1}{n} \dfrac{\partial^2}{\partial \alpha^2} D_{KL}^n(\theta \| \alpha)
\end{cases}
\tag{5.4}
$$

Remark: The interchange of differentiation and integration can be justified by bounded convergence theorem for appropriately well-behaved PDF $f_\alpha(y^M)$. Similar assumptions can be found in Ref. [220]. A sufficient condition for the permission of interchange between differentiation and limitation is the uniform convergence of the limitation in α.

5.1.2 Relations with Fisher Information

Fisher information is a classical criterion for parameter identifiability [216]. There are close relationships between KLID (KLIDR)-identifiability and Fisher information.

The Fisher information matrix (FIM) for the family of densities $\{f_\theta(y^n),$ $\theta \in \Theta \subset \mathbb{R}^d\}$ is given by:

$$
J_F^n(\theta) = E_\theta \left\{ \left(\frac{\partial}{\partial \theta} \log f_\theta(y^n) \right) \left(\frac{\partial}{\partial \theta} \log f_\theta(y^n) \right)^T \right\}
$$

$$
= -E_\theta \left\{ \frac{\partial^2}{\partial \theta^2} \log f_\theta(y^n) \right\}
$$

(5.5)

As $n \to \infty$, the Fisher information rate matrix (FIRM) is:

$$
\bar{J}_F(\theta) = \lim_{n \to \infty} \frac{1}{n} J_F^n(\theta)
$$

(5.6)

Theorem 5.1 Assume that Θ is an open subset of $\mathbb{R}^d$. Then, $\theta \in \Theta$ will be locally KLID-identifiable if the FIM $J_F^M(\theta)$ is positive definite.

Proof: As Θ is an open subset, an obvious sufficient condition for θ to be locally KLID-identifiable is that $((\partial/\partial\alpha)D_{\mathrm{KL}}^M(\theta\|\alpha))|_{\alpha=\theta} = 0$, and $((\partial^2/\partial\alpha^2)D_{\mathrm{KL}}^M(\theta\|\alpha))|_{\alpha=\theta} > 0$. This can be easily proved. By Assumption 5.2, we have

$$
\left(\frac{\partial}{\partial \alpha} D_{\mathrm{KL}}^M(\theta\|\alpha) \right) \Bigg|_{\alpha=\theta} = \left(\frac{\partial}{\partial \alpha} \int f_\theta(y^M) \log \frac{f_\theta(y^M)}{f_\alpha(y^M)} dy^M \right) \Bigg|_{\alpha=\theta}
$$

$$
= -\left(\frac{\partial}{\partial \alpha} \int f_\theta(y^M) \log f_\alpha(y^M) dy^M \right) \Bigg|_{\alpha=\theta}
$$

$$
= -\int \left(f_\theta(y^M) \frac{\frac{\partial}{\partial \alpha} f_\alpha(y^M)}{f_\alpha(y^M)} \right) \Bigg|_{\alpha=\theta} dy^M = -\int \left(\frac{\partial}{\partial \alpha} f_\alpha(y^M) \right) \Bigg|_{\alpha=\theta} dy^M
$$

$$
= -\left(\frac{\partial}{\partial \alpha} \int f_\alpha(y^M) dy^M \right) \Bigg|_{\alpha=\theta} = -\left(\frac{\partial}{\partial \alpha} 1 \right) \Bigg|_{\alpha=\theta} = 0
$$

(5.7)

On the other hand, we can derive

$$
\left(\frac{\partial^2}{\partial \alpha^2} D_{KL}^M(\theta \| \alpha)\right)\Big|_{\alpha=\theta} = \left(\frac{\partial^2}{\partial \alpha^2} \int f_\theta(y^M) \log \frac{f_\theta(y^M)}{f_\alpha(y^M)} dy^M\right)\Big|_{\alpha=\theta}
$$

$$
= -\left(\frac{\partial^2}{\partial \alpha^2} \int f_\theta(y^M) \log f_\alpha(y^M) dy^M\right)\Big|_{\alpha=\theta}
$$

$$
= -\left(\int f_\theta(y^M) \frac{\partial^2 \log f_\alpha(y^M)}{\partial \alpha^2} dy^M\right)\Big|_{\alpha=\theta}
$$

$$
= -E_\theta\left\{\frac{\partial^2 \log f_\theta(y^M)}{\partial \theta^2}\right\} = J_F^M(\theta) > \mathbf{0}
$$

(5.8)

Theorem 5.2 Assume that Θ is an open subset of $\mathbb{R}^d$. Then $\theta \in \Theta$ will be locally KLIDR-identifiable if the FIRM $\bar{J}_F(\theta)$ is positive definite.

Proof: By Theorem 5.1 and Assumption 5.2, we have

$$
\left(\frac{\partial}{\partial \alpha} \bar{D}_{KL}(\theta \| \alpha)\right)\Big|_{\alpha=\theta} = \left(\frac{\partial}{\partial \alpha} \lim_{n \to \infty} \frac{1}{n} D_{KL}^n(\theta \| \alpha)\right)\Big|_{\alpha=\theta}
$$

$$
= \lim_{n \to \infty} \frac{1}{n}\left\{\left(\frac{\partial}{\partial \alpha} D_{KL}^n(\theta \| \alpha)\right)\Big|_{\alpha=\theta}\right\} = 0
$$

(5.9)

and

$$
\left(\frac{\partial^2}{\partial \alpha^2} \bar{D}_{KL}(\theta \| \alpha)\right)\Big|_{\alpha=\theta} = \left(\frac{\partial^2}{\partial \alpha^2} \lim_{n \to \infty} \frac{1}{n} D_{KL}^n(\theta \| \alpha)\right)\Big|_{\alpha=\theta}
$$

$$
= \lim_{n \to \infty} \frac{1}{n}\left\{\left(\frac{\partial^2}{\partial \alpha^2} D_{KL}^n(\theta \| \alpha)\right)\Big|_{\alpha=\theta}\right\}
$$

$$
= \lim_{n \to \infty} \frac{1}{n} J_F^n(\theta) = \bar{J}_F(\theta) > \mathbf{0}
$$

(5.10)

Thus, θ is locally KLIDR-identifiable.

Suppose the observation sequence $\{y_k\}_{k=1}^{\infty}$ ($y_k \in \mathbb{R}$) is a stationary zero-mean Gaussian process, with power spectral $S_\theta(\omega)$. According to Theorem 2.7, the spectral expressions of the KLIDR and FIRM are as follows:

$$\bar{D}_{KL}(\theta \| \alpha) = \lim_{n \to \infty} \frac{1}{n} D_{KL}^n(\theta \| \alpha) = \frac{1}{4\pi} \int_{-\pi}^{\pi} \left\{ \log \frac{S_\alpha(\omega)}{S_\theta(\omega)} + \frac{S_\theta(\omega)}{S_\alpha(\omega)} - 1 \right\} d\omega \quad (5.11)$$

$$\bar{J}_F(\theta) = \lim_{n \to \infty} \frac{1}{n} J_F^n(\theta) = \frac{1}{4\pi} \int_{-\pi}^{\pi} \frac{1}{S_\theta^2(\omega)} \left(\frac{\partial S_\theta(\omega)}{\partial \theta} \right) \left(\frac{\partial S_\theta(\omega)}{\partial \theta} \right)^T d\omega \quad (5.12)$$

In this case, we can easily verify that $((\partial/\partial\alpha)\bar{D}_{KL}(\theta \| \alpha))|_{\alpha=\theta} = 0$, and $((\partial^2/\partial\alpha^2)\bar{D}_{KL}(\theta \| \alpha))|_{\alpha=\theta} = \bar{J}_F(\theta)$. In fact, we have

$$\left(\frac{\partial}{\partial\alpha} \bar{D}_{KL}(\theta \| \alpha) \right) \Bigg|_{\alpha=\theta} = \frac{1}{4\pi} \left(\int_{-\pi}^{\pi} \frac{\partial}{\partial\alpha} \left\{ \log \frac{S_\alpha(\omega)}{S_\theta(\omega)} + \frac{S_\theta(\omega)}{S_\alpha(\omega)} - 1 \right\} d\omega \right) \Bigg|_{\alpha=\theta}$$

$$= \frac{1}{4\pi} \left(\int_{-\pi}^{\pi} \left\{ \left(\frac{S_\alpha(\omega) - S_\theta(\omega)}{S_\alpha^2(\omega)} \right) \frac{\partial}{\partial\alpha} S_\alpha(\omega) \right\} d\omega \right) \Bigg|_{\alpha=\theta}$$

$$= 0$$

$$(5.13)$$

and

$$\left(\frac{\partial^2}{\partial\alpha^2} \bar{D}_{KL}(\theta \| \alpha) \right) \Bigg|_{\alpha=\theta}$$

$$= \frac{1}{4\pi} \frac{\partial}{\partial\alpha} \left(\int_{-\pi}^{\pi} \left\{ \left(\frac{S_\alpha(\omega) - S_\theta(\omega)}{S_\alpha^2(\omega)} \right) \left(\frac{\partial}{\partial\alpha} S_\alpha(\omega) \right)^T \right\} d\omega \right) \Bigg|_{\alpha=\theta}$$

$$= \frac{1}{4\pi} \left(\int_{-\pi}^{\pi} \left\{ [S_\alpha(\omega) - S_\theta(\omega)] \times T(\omega, \alpha) + \frac{1}{S_\alpha^2(\omega)} \left(\frac{\partial}{\partial\alpha} S_\alpha(\omega) \right) \left(\frac{\partial}{\partial\alpha} S_\alpha(\omega) \right)^T \right\} d\omega \right) \Bigg|_{\alpha=\theta}$$

$$= \frac{1}{4\pi} \int_{-\pi}^{\pi} \left\{ \frac{1}{S_\theta^2(\omega)} \left(\frac{\partial}{\partial\theta} S_\theta(\omega) \right) \left(\frac{\partial}{\partial\theta} S_\theta(\omega) \right)^T \right\} d\omega = \bar{J}_F(\theta) \quad (5.14)$$

where $T(\omega, \alpha) = \frac{1}{S_\alpha^3(\omega)} \left(S_\alpha(\omega) \frac{\partial^2}{\partial\alpha^2} S_\alpha(\omega) - 2 \left(\frac{\partial}{\partial\alpha} S_\alpha(\omega) \right) \left(\frac{\partial}{\partial\alpha} S_\alpha(\omega) \right)^T \right)$.

Remark: Theorems 5.1 and 5.2 indicate that, under certain conditions, the positive definiteness of the FIM (or FIRM) provides a sufficient condition for the local KLID (or local KLIDR)-identifiability.

5.1.3 Gaussian Process Case

When the observation sequence $\{y_k\}_{k=1}^{\infty}$ is jointly Gaussian distributed, the KLID-identifiability can be easily checked. Consider the following joint Gaussian PDF:

$$f_\theta(y^n) = \frac{1}{(2\pi)^{mn/2}\sqrt{\det \Delta_\theta^n}} \exp\left\{-\frac{1}{2}(y^n - \bar{y}_\theta^n)^T(\Delta_\theta^n)^{-1}(y^n - \bar{y}_\theta^n)\right\} \tag{5.15}$$

where $\bar{y}_\theta^n = E_\theta[y^n]$ is the mean vector, and $\Delta_\theta^n = E_\theta[(y^n - \bar{y}_\theta^n)(y^n - \bar{y}_\theta^n)^T]$ is the $mn \times mn$-dimensional covariance matrix. Then we have

$$D_{KL}^n(\theta\|\alpha) = \tilde{D}_{KL}^n(\theta\|\alpha) + \frac{1}{2}(\bar{y}_\theta^n - \bar{y}_\alpha^n)^T(\Delta_\alpha^n)^{-1}(\bar{y}_\theta^n - \bar{y}_\alpha^n) \tag{5.16}$$

where $\tilde{D}_{KL}^n(\theta\|\alpha)$ is

$$\tilde{D}_{KL}^n(\theta\|\alpha) = \frac{1}{2}\left\{\log\frac{|\Delta_\alpha^n|}{|\Delta_\theta^n|} + Tr(\Delta_\theta^n((\Delta_\alpha^n)^{-1} - (\Delta_\theta^n)^{-1}))\right\} \tag{5.17}$$

Clearly, for the Gaussian process $\{y_k\}_{k=1}^{\infty}$, we have $D_{KL}^M(\theta\|\alpha) = 0$ if and only if $\Delta_\theta^M = \Delta_\alpha^M$ and $\bar{y}_\theta^M = \bar{y}_\alpha^M$. Denote $\Delta_\theta^M = ((\Delta_\theta^M)_{ij})$, $i,j = 1, 2, \ldots, mM$, where $(\Delta_\theta^M)_{ij}$ is the ith row and jth column element of Δ_θ^M. The element $(\Delta_\theta^M)_{ij}$ is said to be a regular element if and only if $(\partial/\partial\theta)(\Delta_\theta^M)_{ij} \not\equiv 0$, i.e., as a function of θ, $(\Delta_\theta^M)_{ij}$ is not a constant. In a similar way, we define the regular element of the mean vector $\bar{y}_\theta^M$. Let $\Psi_M(\theta)$ be a column vector containing all the distinct regular elements from Δ_θ^M and $\bar{y}_\theta^M$. We call $\Psi_M(\theta)$ the regular characteristic vector (RCV) of the Gaussian process $\{y_k\}_{k=1}^{\infty}$. Then we have $\Psi_M(\theta) = \Psi_M(\alpha)$ if and only if $D_{KL}^M(\theta\|\alpha) = 0$. According to Definition 5.1, for the Gaussian process $\{y_k\}_{k=1}^{\infty}$, the parameter set Θ is KLID-identifiable at $\theta \in \Theta$, if and only if $\exists M \in \mathbb{N}$, $\forall \alpha \in \Theta$, $\Psi_M(\theta) = \Psi_M(\alpha)$ implies $\alpha = \theta$.

Assume that Θ is an open subset of $\mathbb{R}^d$. By Lemma 1 of Ref. [219], the map $\theta \mapsto \Psi_M(\theta)$ will be locally one to one at $\theta = \tilde{\theta}$ if the Jacobian of $\Psi_M(\theta)$ has full rank d at $\theta = \tilde{\theta}$. Therefore, a sufficient condition for θ to be locally KLID-identifiable is that

$$\text{rank}\left(\frac{\partial}{\partial\theta^T}\Psi_M(\theta)\right) = d \tag{5.18}$$

Example 5.1 Consider the following second-order state-space model ($m = 1, d = 2$) [120]:

$$\begin{cases} \begin{pmatrix} x_{1,k+1} \\ x_{2,k+1} \end{pmatrix} = \begin{pmatrix} \theta_1 & 0 \\ 1 & \theta_2 \end{pmatrix} \begin{pmatrix} x_{1,k} \\ x_{2,k} \end{pmatrix} + \begin{pmatrix} 1 \\ 0 \end{pmatrix} w_k, \quad \begin{pmatrix} x_{1,0} \\ x_{2,0} \end{pmatrix} = \begin{pmatrix} 0 \\ 0 \end{pmatrix}, \quad \theta \in \mathbb{R}^2 \\ y_k = x_{2,k} \end{cases}$$

$$(5.19)$$

where $\{w_k\}$ is a zero-mean white Gaussian process with unit power. Then the output sequence with $M = 4$ is

$$y^4 = \begin{pmatrix} y_1 \\ y_2 \\ y_3 \\ y_4 \end{pmatrix} = \begin{pmatrix} 0 \\ w_0 \\ (\theta_1 + \theta_2)w_0 + w_1 \\ (\theta_1^2 + \theta_1\theta_2 + \theta_2^2)w_0 + (\theta_1 + \theta_2)w_1 + w_2 \end{pmatrix}$$

$$(5.20)$$

It is easy to obtain the RCV:

$$\Psi_4(\theta) = \begin{pmatrix} \theta_1 + \theta_2 \\ \theta_1^2 + \theta_1\theta_2 + \theta_2^2 \\ (\theta_1 + \theta_2)^2 + 1 \\ (\theta_1 + \theta_2)(\theta_1^2 + \theta_1\theta_2 + \theta_2^2 + 1) \\ (\theta_1^2 + \theta_1\theta_2 + \theta_2^2)^2 + (\theta_1 + \theta_2)^2 + 1 \end{pmatrix}$$

$$(5.21)$$

The Jacobian matrix can then be calculated as:

$$\frac{\partial}{\partial \theta^T} \Psi_4(\theta)$$

$$= \begin{pmatrix} 1 & 1 \\ 2\theta_1 + \theta_2 & \theta_1 + 2\theta_2 \\ 2(\theta_1 + \theta_2) & 2(\theta_1 + \theta_2) \\ 3\theta_1^2 + 4\theta_1\theta_2 + 2\theta_2^2 + 1 & 2\theta_1^2 + 4\theta_1\theta_2 + 3\theta_2^2 + 1 \\ 2(2\theta_1^3 + 3\theta_1^2\theta_2 + 3\theta_1\theta_2^2 + \theta_2^3 + \theta_1 + \theta_2) & 2(\theta_1^3 + 3\theta_1^2\theta_2 + 3\theta_1\theta_2^2 + 2\theta_2^3 + \theta_1 + \theta_2) \end{pmatrix}$$

$$(5.22)$$

Clearly, we have $\operatorname{rank}\left(\dfrac{\partial}{\partial \theta^T}\Psi_4(\theta)\right) = 2$ for all $\theta \in \mathbb{R}^2$ with $\theta_1 \neq \theta_2$. So this parameterization is locally KLID-identifiable provided $\theta_1 \neq \theta_2$. The identifiability can also be checked from the transfer function. The transfer function of the above system is:

$$G_\theta(z) = \frac{1}{(z - \theta_1)(z - \theta_2)} \tag{5.23}$$

$\forall \theta \in \mathbb{R}^2$, $\theta_1 \neq \theta_2$, define $\varepsilon = (1/2)|\theta_1 - \theta_2|$. Then $\forall \alpha, \beta \in \mathbb{R}^2$, we have $\alpha = \beta$ provided the following two conditions are met:

1. $\|\alpha - \theta\| < \varepsilon$, $\|\beta - \theta\| < \varepsilon$
2. $\forall z \in \mathbb{C}$, $z \neq \theta_1, \theta_2$, $G_\alpha(z) = G_\beta(z)$

According to the Definition 1 of Ref. [219], this system is also locally identifiable from the transfer function provided $\theta_1 \neq \theta_2$.

The KLID-identifiability also has connection with the LS-identifiability [217]. Consider the signal-plus-noise model:

$$z_\theta(k) = y_\theta(k) + v_k, \quad y_\theta(k) \in \mathbb{R}^m \tag{5.24}$$

where $\{y_\theta(k)\}$ is a parameterized deterministic signal, $\{v_k\}$ is a zero-mean white Gaussian noise, $E[v_i v_j^T] = I\delta_{ij}$ (I is an $m \times m$-dimensional identity matrix), and $\{z_\theta(k)\}$ is the noisy observation. Then we have

$$\begin{cases} E_\theta(z_\theta(k)) = E_\theta(y_\theta(k) + v_k) = y_\theta(k) \\ E_\theta[(z_\theta(i) - E_\theta(z_\theta(i)))(z_\theta(j) - E_\theta(z_\theta(j)))^T] = E[v_i v_j^T] = I\delta_{ij} \end{cases} \tag{5.25}$$

By Eq. (5.16), we derive

$$D_{KL}^M(\theta \| \alpha) = D_{KL}(f_\theta(z^M) \| f_\alpha(z^M))$$

$$= D_{KL}^M(v^M \| v^M) + \frac{1}{2}(y_\theta^M - y_\alpha^M)^T (\Delta^M)^{-1}(y_\theta^M - y_\alpha^M) \tag{5.26}$$

$$= \frac{1}{2} \sum_{i=1}^{M} \|y_\theta(i) - y_\alpha(i)\|^2$$

where $\Delta^M = \mathrm{diag}[I, I, \ldots, I]$. The above KLID is equivalent to the LS criterion of the deterministic part. In this case, the KLID-identifiability reduces to the LS-identifiability of the deterministic part.

Next, we show that for a stationary Gaussian process, the KLIDR-identifiability is identical to the identifiability from the output spectral density [219].

Let $\{y_\theta(k) \in \mathbb{R}^m\}_{k=1}^{\infty}$ ($\theta \in \Theta$) be a parameterized zero-mean stationary Gaussian process with continuous spectral density $S_\theta(\omega)$ ($m \times m$-dimensional matrix).

By Theorem 2.7, the KLIDR between $\{y_\theta(k)\}_{k=1}^\infty$ and $\{y_\alpha(k)\}_{k=1}^\infty$ exists and is given by:

$$
\begin{aligned}
\overline{D}_{\mathrm{KL}}(\theta\|\alpha) &= \lim_{n\to\infty} \frac{1}{n} D_{\mathrm{KL}}^n(\theta\|\alpha) \\
&= \frac{1}{4\pi}\int_{-\pi}^{\pi} \left\{ \log\frac{\det S_\alpha(\omega)}{\det S_\theta(\omega)} + Tr(S_\alpha(\omega)^{-1}[S_\theta(\omega) - S_\alpha(\omega)]) \right\} d\omega
\end{aligned}
\tag{5.27}
$$

Theorem 5.3 $\overline{D}_{\mathrm{KL}}(\theta\|\alpha) \geq 0$, with equality if and only if $S_\alpha(\omega) = S_\theta(\omega)$, $\forall\,\omega\in\mathbb{R}$.

Proof: $\forall\,\omega\in\mathbb{R}$, the spectral density matrices $S_\theta(\omega)$ and $S_\alpha(\omega)$ are positive definite. Let $X_\theta(\omega)$ and $X_\alpha(\omega)$ be two normally distributed m-dimensional vectors, $X_\theta(\omega) \sim \mathcal{N}(0, S_\theta(\omega))$ and $X_\alpha(\omega) \sim \mathcal{N}(0, S_\alpha(\omega))$, respectively. Then we have

$$
D_{\mathrm{KL}}(X_\theta(\omega)\|X_\alpha(\omega)) = \frac{1}{2}\left\{ \log\frac{\det S_\alpha(\omega)}{\det S_\theta(\omega)} + Tr(S_\alpha(\omega)^{-1}[S_\theta(\omega) - S_\alpha(\omega)]) \right\}
\tag{5.28}
$$

Combining Eqs. (5.28) and (5.27) yields

$$
\overline{D}_{\mathrm{KL}}(\theta\|\alpha) = \frac{1}{2\pi}\int_{-\pi}^{\pi} D_{\mathrm{KL}}(X_\theta(\omega)\|X_\alpha(\omega)) d\omega
\tag{5.29}
$$

It follows easily that $\overline{D}_{\mathrm{KL}}(\theta\|\alpha) \geq 0$, with equality if and only if $D_{\mathrm{KL}}(X_\theta(\omega)\|X_\alpha(\omega)) = 0$ for almost every ω (hence $S_\alpha(\omega) = S_\theta(\omega)$, $\forall\,\omega\in\mathbb{R}$).

By Theorem 5.3, we may conclude that for a stationary Gaussian process, θ is KLIDR-identifiable if and only if $\forall\,\alpha\in\Theta$, $S_\alpha(\omega) = S_\theta(\omega)$ implies $\alpha = \theta$. This is exactly the identifiability from the output spectral density.

5.1.4 Markov Process Case

Now we focus on situations where the observation sequence is a parameterized Markov process. First, let us define the minimum identifiable horizon (MIH).

Definition 5.4 Assume that $\theta\in\Theta$ is KLID-identifiable. Then the MIH is [120]:

$$
\mathrm{MIH}(\theta) = \min \mathbf{M}_\theta
\tag{5.30}
$$

where $\mathbf{M}_\theta = \{M | M \in \mathbb{N}, \theta \text{ is KLID-identifiable over}[1, M]\}$.

The MIH is the minimum length of the observation sequence from which θ can be uniquely identified. If the MIH is known, we could identify θ with the least observation data. In general, it is difficult to obtain the exact value of MIH. In some special situations, however, one can derive an upper bound on the MIH. For a parameterized Markov process, this upper bound is straightforward. In the theorem below, we show that for a $(p-1)$-order strictly stationary Markov process, the number p provides an upper bound on the MIH.

Theorem 5.4 If the observation sequence $\{y_\theta(k)\}_{k=1}^\infty$ $(\theta \in \Theta)$ is a $(p-1)$-order strictly stationary Markov process $(p \geq 1)$, and the parameter set Θ is KLID-identifiable at $\theta \in \Theta$, then we have MIH$(\theta) \leq p$.

Proof: As parameter set Θ is KLID-identifiable at θ, by Definition 5.1, there exists a number $M \in \mathbf{M}_\theta \subset \mathbb{N}$, such that $\forall \alpha \in \Theta$, $D_{KL}^M(\theta \| \alpha) = 0 \Leftrightarrow \alpha = \theta$. Let us consider two cases, one for which $p = 1$ and the other for which $p > 1$.

1. $p = 1$: The zero-order strictly stationary Markov process refers to an independent and identically distributed sequence. In this case, we have $f_\theta(y^M) = \prod_{i=1}^M f_\theta(y_i)$, and

$$
D_{KL}^M(\theta \| \alpha) = \int f_\theta(y^M) \log \frac{f_\theta(y^M)}{f_\alpha(y^M)} \, dy^M
$$

$$
= \int \prod_{i=1}^M f_\theta(y_i) \log \frac{\prod_{i=1}^M f_\theta(y_i)}{\prod_{i=1}^M f_\alpha(y_i)} \, dy^M \tag{5.31}
$$

$$
= \sum_{i=1}^M \int f_\theta(y_i) \log \frac{f_\theta(y_i)}{f_\alpha(y_i)} \, dy_i = M D_{KL}^1(\theta \| \alpha)
$$

And hence, $\forall \alpha \in \Theta$, we have $D_{KL}^1(\theta \| \alpha) = 0 \Leftrightarrow D_{KL}^M(\theta \| \alpha) = 0 \Leftrightarrow \alpha = \theta$. It follows that $1 \in \mathbf{M}_\theta$, and MIH$(\theta) = \min \mathbf{M}_\theta = 1 \leq p$.

2. $p > 1$: If $p \geq M$, then MIH$(\theta) = \min \mathbf{M}_\theta \leq M \leq p$. If $1 < p < M$, then

$$
f_\theta(y^M) = f_\theta(y^{p-1}) \prod_{i=p}^M f_\theta(y_i | y^{i-1}) \tag{5.32}
$$

By Markovian and stationary properties, one can derive

$$
\begin{aligned}
f_\theta(y^M) &= f_\theta(y^{p-1}) \prod_{i=p}^{M} f_\theta(y_i|y^{i-1}) \\
&= f_\theta(y^{p-1}) \prod_{i=p}^{M} f_\theta(y_i|y_1,\ldots,y_{i-1}) \\
&= f_\theta(y^{p-1}) \prod_{i=p}^{M} f_\theta(y_i|y_{i-p+1},\ldots,y_{i-1}) \\
&= f_\theta(y^{p-1}) \prod_{i=p}^{M} f_\theta(y_p|y_1,\ldots,y_{p-1}) \\
&= f_\theta(y^{p-1}) \prod_{i=p}^{M} f_\theta(y_p|y^{p-1})
\end{aligned}
\tag{5.33}
$$

It follows that

$$
\begin{aligned}
D_{KL}^M(\theta\|\alpha) &= \int f_\theta(y^M)\log \frac{f_\theta(y^{p-1})\prod_{i=p}^{M} f_\theta(y_p|y^{p-1})}{f_\alpha(y^{p-1})\prod_{i=p}^{M} f_\alpha(y_p|y^{p-1})} dy^M \\
&= \int f_\theta(y^{p-1})\log\frac{f_\theta(y^{p-1})}{f_\alpha(y^{p-1})} dy^{p-1} + M\int f_\theta(y^p)\log\frac{f_\theta(y_p|y^{p-1})}{f_\alpha(y_p|y^{p-1})} dy^p \\
&= D_{KL}^{p-1}(\theta\|\alpha) + M D_{KL}(f_\theta(y_p|y^{p-1})\|f_\alpha(y_p|y^{p-1}))
\end{aligned}
\tag{5.34}
$$

where $D_{KL}(f_\theta(y_p|y^{p-1})\|f_\alpha(y_p|y^{p-1}))$ is the conditional KLID. And hence,

$$
D_{KL}^M(\theta\|\alpha) = 0 \Leftrightarrow \begin{cases} D_{KL}^{p-1}(\theta\|\alpha) = 0 \\ D_{KL}(f_\theta(y_p|y^{p-1})\|f_\alpha(y_p|y^{p-1})) = 0 \end{cases}
$$

$$
\Leftrightarrow \begin{cases} f_\theta(y^{p-1}) = f_\alpha(y^{p-1}) \\ f_\theta(y_p|y^{p-1}) = f_\alpha(y_p|y^{p-1}) \end{cases}
$$

$$
\Leftrightarrow f_\theta(y^p) = f_\alpha(y^p)
$$

$$
\Leftrightarrow D_{KL}^p(\theta\|\alpha) = 0
\tag{5.35}
$$

Then $\forall\,\alpha\in\Theta$, we have $D_{KL}^p(\theta\|\alpha) = 0 \Leftrightarrow D_{KL}^M(\theta\|\alpha) = 0 \Leftrightarrow \alpha = \theta$. Thus, $p\in\mathbf{M}_\theta$, and it follows that $\mathrm{MIH}(\theta) = \min\mathbf{M}_\theta \le p$.

Example 5.2 Consider the first-order AR model $(d = 2)$ [120]:

$$y_k = \theta_1 y_{k-1} + \theta_2 \nu_k, 0 < \theta_1 < 1, \theta_2 \neq 0 \tag{5.36}$$

where $\{\nu_k\}$ is a zero-mean white Gaussian noise with unit power. Assume that the system has reached steady state when the observations begin. The observation sequence $\{y_k\}$ will be a first-order stationary Gaussian Markov process, with covariance matrix:

$$\Delta_\theta^n = \frac{\theta_2^2}{1 - \theta_1^2} \begin{pmatrix} 1 & \theta_1 & \theta_1^2 & \theta_1^3 & \cdots & \theta_1^{n-1} \\ \theta_1 & 1 & \theta_1 & \theta_1^2 & \cdots & \theta_1^{n-2} \\ \theta_1^2 & \theta_1 & 1 & \theta_1 & \cdots & \theta_1^{n-3} \\ \theta_1^3 & \theta_1^2 & \theta_1 & 1 & \cdots & \theta_1^{n-4} \\ \vdots & \vdots & \vdots & \vdots & \ddots & \vdots \\ \theta_1^{n-1} & \theta_1^{n-2} & \theta_1^{n-3} & \theta_1^{n-4} & \cdots & 1 \end{pmatrix}_{n \times n} \tag{5.37}$$

$\forall\, \theta \in \Theta$ $(0 < \theta_1 < 1,\, \theta_2 \neq 0)$, we have

$$\text{rank}\left\{ \frac{\partial}{\partial \theta^T} \Psi_2(\theta) \right\} = \text{rank}\left\{ \frac{\partial}{\partial \theta^T} \Psi_3(\theta) \right\} = \cdots = \text{rank}\left\{ \frac{\partial}{\partial \theta^T} \Psi_n(\theta) \right\} = 2 = d \tag{5.38}$$

where $\Psi_i(\theta)$ are the RCVs. And hence, $\text{MIH}(\theta) = 2 \leq p$.

The following corollary is a direct consequence of Theorem 5.4.

Corollary 5.1 For a $(p - 1)$-order strictly stationary Markov process $\{y_\theta(k)\}_{k=1}^\infty$, the parameter set Θ is KLID-identifiable at $\theta \in \Theta$ if and only if $\forall\, \alpha \in \Theta$, $D_{\text{KL}}^p(\theta \| \alpha) = 0$ implies $\alpha = \theta$.

From the theory of stochastic process, for a $(p - 1)$-order strictly stationary Markov process $\{y_\theta(k)\}_{k=1}^\infty$, under certain conditions (see Ref. [221] for details), the conditional density $f_\theta(y_p | y^{p-1})$ will determine uniquely the joint density $f_\theta(y^p)$. In this case, the KLID-identifiability and the KLIDR-identifiability are equivalent.

Theorem 5.5 Assume that the observation sequence $\{y_\theta(k)\}_{k=1}^\infty$ is a $(p - 1)$-order strictly stationary Markov process $(p \geq 1)$, whose conditional density $f_\theta(y_p | y^{p-1})$ uniquely determines the joint density $f_\theta(y^p)$. Then, $\forall\, \theta \in \Theta$, θ is KLID-identifiable if and only if it is KLIDR-identifiable.

Proof: We only need to prove $\lim\limits_{n \to \infty} (1/n) D_{\text{KL}}^n(\theta \| \alpha) = 0 \Leftrightarrow D_{\text{KL}}^p(\theta \| \alpha) = 0$.

1. When $p = 1$, we have $D_{\text{KL}}^n(\theta \| \alpha) = n D_{\text{KL}}^1(\theta \| \alpha)$, and hence $\lim\limits_{n \to \infty} \frac{1}{n} D_{\text{KL}}^n(\theta \| \alpha) = D_{\text{KL}}^1(\theta \| \alpha)$.

2. When $p > 1$, we have $D_{KL}^n(\theta \| \alpha) = D_{KL}^{p-1}(\theta \| \alpha) + n D_{KL}(f_\theta(y_p | y^{p-1}) \| f_\alpha(y_p | y^{p-1}))$, and it follows that

$$\lim_{n \to \infty} \frac{1}{n} D_{KL}^n(\theta \| \alpha)$$

$$= \lim_{n \to \infty} \left\{ \frac{1}{n} D_{KL}^{p-1}(\theta \| \alpha) + D_{KL}(f_\theta(y_p | y^{p-1}) \| f_\alpha(y_p | y^{p-1})) \right\}$$

$$= D_{KL}(f_\theta(y_p | y^{p-1}) \| f_\alpha(y_p | y^{p-1})) \tag{5.39}$$

Since $f_\theta(y_p | y^{p-1})$ uniquely determines $f_\theta(y^p)$, we can derive

$$\lim_{n \to \infty} \frac{1}{n} D_{KL}^n(\theta \| \alpha) = 0$$

$$\Leftrightarrow D_{KL}(f_\theta(y_p | y^{p-1}) \| f_\alpha(y_p | y^{p-1})) = 0$$

$$\Leftrightarrow f_\theta(y_p | y^{p-1}) = f_\alpha(y_p | y^{p-1})$$

$$\Leftrightarrow f_\theta(y^p) = f_\alpha(y^p)$$

$$\Leftrightarrow D_{KL}^p(\theta \| \alpha) = 0 \tag{5.40}$$

This completes the proof.

5.1.5   Asymptotic KLID-Identifiability

In the previous discussions, we assume that the true density $f_{\theta_0}(y^M)$ is known. In most practical situations, however, the actual density, and hence the KLID, needs to be estimated using random data drawn from the underlying density. Let $(y_{(1)}^M, \ldots, y_{(L)}^M)$ be an independent and identically distributed (i.i.d.) sample drawn from $f_{\theta_0}(y^M)$. The density estimator for $f_{\theta_0}(y^M)$ will be a mapping $\hat{f}_L : \mathbb{R}^{mM} \times (\mathbb{R}^{mM})^L \to \mathbb{R}$ [98]:

$$\begin{cases} \hat{f}_L(y^M) = \hat{f}_L(y^M; y_{(1)}^M, \ldots, y_{(L)}^M) \geq 0 \\ \int \hat{f}_L(y^M) dy^M = 1 \end{cases} \tag{5.41}$$

The asymptotic KLID-identifiability is then defined as follows:

Definition 5.5 The parameter set Θ is said to be asymptotic KLID-identifiable at $\theta_0 \in \Theta$, if there exists a sequence of density estimates $\{\hat{f}_L\}_{L=1}^{\infty}$, such that $\hat{\theta}_L \xrightarrow[L \to \infty]{P} \theta_0$ (convergence in probability), where $\hat{\theta}_L$ is the minimum KLID estimator, $\hat{\theta}_L = \arg\min_{\theta \in \Theta} D_{KL}^M(\hat{f}_L \| f_\theta)$.

Theorem 5.6 Assume that the parameter space Θ is a compact subset, and the density estimate sequence $\{\hat{f}_L\}_{L=1}^{\infty}$ satisfies $D_{\text{KL}}^M(\hat{f}_L\|f_{\theta_0}) \xrightarrow[L\to\infty]{P} 0$. Then, θ_0 will be asymptotic KLID-identifiable provided it is KLID-identifiable.

Proof: Since $D_{\text{KL}}^M(\hat{f}_L\|f_{\theta_0}) \xrightarrow[L\to\infty]{P} 0$, for $\varepsilon > 0$ and $\delta > 0$ arbitrarily small, there exists an $N(\varepsilon, \delta) < \infty$ such that for $L > N(\varepsilon, \delta)$,

$$\Pr\{D_{\text{KL}}^M(\hat{f}_L\|f_{\theta_0}) > \delta\} < \varepsilon \tag{5.42}$$

where $\Pr\{D_{\text{KL}}^M(\hat{f}_L\|f_{\theta_0}) > \delta\}$ is the probability of Borel set $\{\{y_{(1)}^M, \ldots, y_{(L)}^M\}|D_{\text{KL}}^M(\hat{f}_L\|f_{\theta_0}) > \delta\}$. On the other hand, as $\hat{\theta}_L = \arg\min_{\theta \in \Theta} D_{\text{KL}}^M(\hat{f}_L\|f_\theta)$, we have

$$D_{\text{KL}}^M(\hat{f}_L\|f_{\hat{\theta}_L}) = \min_{\theta \in \Theta} D_{\text{KL}}^M(\hat{f}_L\|f_\theta) \leq D_{\text{KL}}^M(\hat{f}_L\|f_{\theta_0}) \tag{5.43}$$

Then the event $\{D_{\text{KL}}^M(\hat{f}_L\|f_{\hat{\theta}_L}) > \delta\} \subset \{D_{\text{KL}}^M(\hat{f}_L\|f_{\theta_0}) > \delta\}$, and hence

$$\Pr\{D_{\text{KL}}^M(\hat{f}_L\|f_{\hat{\theta}_L}) > \delta\} \leq \Pr\{D_{\text{KL}}^M(\hat{f}_L\|f_{\theta_0}) > \delta\} < \varepsilon \tag{5.44}$$

By Pinsker's inequality, we have

$$\begin{cases} D_{\text{KL}}^M(\hat{f}_L\|f_{\theta_0}) \geq \dfrac{1}{2}\|\hat{f}_L - f_{\theta_0}\|_1^2 \\[2mm] D_{\text{KL}}^M(\hat{f}_L\|f_{\hat{\theta}_L}) \geq \dfrac{1}{2}\|\hat{f}_L - f_{\hat{\theta}_L}\|_1^2 \end{cases} \tag{5.45}$$

where $\|\hat{f}_L - f_{\hat{\theta}_L}\|_1 = \int |\hat{f}_L - f_{\hat{\theta}_L}| dy^M$ is the L_1-distance (or the total variation). It follows that

$$\begin{cases} \Pr\left\{\|\hat{f}_L - f_{\theta_0}\|_1 > \sqrt{2\delta}\right\} \leq \Pr\{D_{\text{KL}}^M(\hat{f}_L\|f_{\theta_0}) > \delta\} < \varepsilon \\[2mm] \Pr\left\{\|\hat{f}_L - f_{\hat{\theta}_L}\|_1 > \sqrt{2\delta}\right\} \leq \Pr\{D_{\text{KL}}^M(\hat{f}_L\|f_{\hat{\theta}_L}) > \delta\} < \varepsilon \end{cases} \tag{5.46}$$

In addition, the following inequality holds:

$$\|f_{\hat{\theta}_L} - f_{\theta_0}\|_1 = \|(\hat{f}_L - f_{\theta_0}) - (\hat{f}_L - f_{\hat{\theta}_L})\|_1 \leq \|\hat{f}_L - f_{\theta_0}\|_1 + \|\hat{f}_L - f_{\hat{\theta}_L}\|_1 \tag{5.47}$$

Then we have

$$\left\{\|f_{\hat{\theta}_L} - f_{\theta_0}\|_1 > 2\sqrt{2\delta}\right\} \subset \left(\left\{\|\hat{f}_L - f_{\theta_0}\|_1 > \sqrt{2\delta}\right\} \cup \left\{\|\hat{f}_L - f_{\hat{\theta}_L}\|_1 > \sqrt{2\delta}\right\}\right) \tag{5.48}$$

And hence

$$\Pr\left\{\left\|f_{\hat{\theta}_L}-f_{\theta_0}\right\|_1>2\sqrt{2\delta}\right\}$$

$$\leq\Pr\left(\left\{\left\|\hat{f}_L-f_{\theta_0}\right\|_1>\sqrt{2\delta}\right\}\cup\left\{\left\|\hat{f}_L-f_{\hat{\theta}_L}\right\|_1>\sqrt{2\delta}\right\}\right)$$

$$\leq\Pr\left\{\left\|\hat{f}_L-f_{\theta_0}\right\|_1>\sqrt{2\delta}\right\}+\Pr\left\{\left\|\hat{f}_L-f_{\hat{\theta}_L}\right\|_1>\sqrt{2\delta}\right\}$$

$$<2\varepsilon$$

(5.49)

For any $\tau>0$, we define the set $\Theta_\tau=\{\alpha|\alpha\in\Theta,\|\alpha-\theta_0\|\geq\tau\}$, where $\|\cdot\|$ is the Euclidean norm. As Θ is a compact subset in $\mathbb{R}^d$, Θ_τ must be a compact set too. Meanwhile, by Assumption 5.2, the function $\varphi(\alpha)=\left\|f_\alpha-f_{\theta_0}\right\|_1$ ($\alpha\in\Theta$) will be a continuous mapping $\varphi:\Theta\to\mathbb{R}$. Thus, a minimum of $\varphi(\alpha)$ over the set Θ_τ must exist. Denote $\gamma_\tau=\min_{\alpha\in\Theta_\tau}\left\|f_\alpha-f_{\theta_0}\right\|_1$, it follows easily that

$$\left\{\left\|\hat{\theta}_L-\theta_0\right\|\geq\tau\right\}\subset\left\{\left\|f_{\hat{\theta}_L}-f_{\theta_0}\right\|_1\geq\gamma_\tau\right\}$$

(5.50)

If θ_0 is KLID-identifiable, $\forall\alpha\in\Theta$, $\alpha\neq\theta_0$, we have $D_{\mathrm{KL}}^M(\theta\|\alpha)\neq0$, or equivalently, $\left\|f_\alpha-f_{\theta_0}\right\|_1\neq0$. It follows that $\gamma_\tau=\min_{\alpha\in\Theta_\tau}\left\|f_\alpha-f_{\theta_0}\right\|_1\neq0$. Let $\delta=\frac{1}{32}\gamma_\tau^2>0$, we have

$$\Pr\{\left\|\hat{\theta}_L-\theta_0\right\|>\tau\}\leq\Pr\{\left\|\hat{\theta}_L-\theta_0\right\|\geq\tau\}$$

$$\leq\Pr\{\left\|f_{\hat{\theta}_L}-f_{\theta_0}\right\|_1\geq\gamma_\tau\}$$

$$\leq\Pr\left\{\left\|f_{\hat{\theta}_L}-f_{\theta_0}\right\|_1>\frac{1}{2}\gamma_\tau\}\right.$$

$$=\Pr\left\{\left\|f_{\hat{\theta}_L}-f_{\theta_0}\right\|_1>2\sqrt{2\delta}\right\}<2\varepsilon$$

(5.51)

This implies $\left\|\hat{\theta}_L-\theta_0\right\|\xrightarrow[L\to\infty]{P}0$, and hence $\hat{\theta}_L\xrightarrow[L\to\infty]{P}\theta_0$.

According to Theorem 5.6, if the density estimate $\hat{f}_L$ is consistent in KLID in probability ($D_{\mathrm{KL}}^M(\hat{f}_L\|f_{\theta_0})\xrightarrow[L\to\infty]{P}0$), the KLID-identifiability will be a sufficient condition for the asymptotic KLID-identifiability. The next theorem shows that, under certain conditions, the KLID-identifiability will also be a necessary condition for θ_0 to be asymptotic KLID-identifiable.

Theorem 5.7 If $\theta_0\in\Theta$ is asymptotic KLID-identifiable, then it is KLID-identifiable provided

1. Θ is a compact subset in $\mathbb{R}^d$,
2. $\forall\alpha\in\Theta$, if $D_{\mathrm{KL}}^M(\theta_0\|\alpha)=0$, then there exist $\varepsilon>0$ and an infinite set $S\subset\mathbb{N}$ such that for $L\in S$,

$$\Pr\{\min_{\theta \in \overline{B}(\alpha,\kappa)} D_{\mathrm{KL}}^M(\hat{f}_L \| f_\theta) \leq \min_{\theta \in \overline{B}(\theta_0,\kappa)} D_{\mathrm{KL}}^M(\hat{f}_L \| f_\theta)\} > \varepsilon \qquad (5.52)$$

where $\overline{B}(\alpha,\kappa) = \{x \mid x \in \Theta, \|x - \alpha\| \leq \kappa\}$, $\kappa = \frac{1}{3}\|\alpha - \theta_0\|$.

Proof: If θ_0 is asymptotic KLID-identifiable, then for $\varepsilon > 0$ and $\delta > 0$ arbitrarily small, there exists an $N(\varepsilon, \delta) < \infty$ such that for $L > N(\varepsilon, \delta)$,

$$\Pr\{\|\hat{\theta}_L - \theta_0\| > \delta\} < \varepsilon \qquad (5.53)$$

Suppose θ_0 is not KLID-identifiable, then $\exists \alpha \in \Theta$, $\alpha \neq \theta_0$, such that $D_{\mathrm{KL}}^M(\theta_0 \| \alpha) = 0$. Let $\delta = \kappa = (1/3)\|\alpha - \theta_0\| > 0$, we have (as Θ is a compact subset, the minimum exists)

$$\Pr\{\|\hat{\theta}_L - \theta_0\| > \kappa\} < \varepsilon \Rightarrow \Pr\{\hat{\theta}_L \notin \overline{B}(\theta_0, \kappa)\} < \varepsilon$$

$$\Rightarrow \Pr\{\arg \min_{\theta \in \Theta} D_{\mathrm{KL}}^M(\hat{f}_L \| f_\theta) \notin \overline{B}(\theta_0, \kappa)\} < \varepsilon$$

$$\Rightarrow \Pr\{\min_{\theta \in \{\Theta - \overline{B}(\theta_0,\kappa)\}} D_{\mathrm{KL}}^M(\hat{f}_L \| f_\theta) \leq \min_{\theta \in \overline{B}(\theta_0,\kappa)} D_{\mathrm{KL}}^M(\hat{f}_L \| f_\theta)\} < \varepsilon \qquad (5.54)$$

$$\stackrel{(a)}{\Rightarrow} \Pr\{\min_{\theta \in \overline{B}(\alpha,\kappa)} D_{\mathrm{KL}}^M(\hat{f}_L \| f_\theta) \leq \min_{\theta \in \overline{B}(\theta_0,\kappa)} D_{\mathrm{KL}}^M(\hat{f}_L \| f_\theta)\} < \varepsilon$$

where (a) follows from $\overline{B}(\alpha, \kappa) \subset \{\Theta - \overline{B}(\theta_0, \kappa)\}$. The above result contradicts the condition (2). Therefore, θ_0 must be KLID-identifiable.

In the following, we consider several specific density estimation methods and discuss the consistency problems of the related parameter estimators.

5.1.5.1 Maximum Likelihood Estimation

The maximum likelihood estimation (MLE) is a popular parameter estimation method and is also an important parametric approach for the density estimation. By MLE, the density estimator is

$$\hat{f}_L(y^M) = f_{\hat{\theta}_{\mathrm{ML}}}(y^M) \qquad (5.55)$$

where $\hat{\theta}_{\mathrm{ML}} \in \Theta$ is obtained by maximizing the likelihood function, that is,

$$\hat{\theta}_{\mathrm{ML}} = \arg \max_{\theta \in \Theta} \prod_{i=1}^{L} f_\theta(y_{(i)}^M) \qquad (5.56)$$

Lemma 5.1 The MLE density estimate sequence $\{f_{\hat{\theta}_{ML}}(y^M)\}_{L=1}^{\infty}$ satisfies $D_{KL}^M(f_{\hat{\theta}_{ML}} \| f_{\theta_0}) \xrightarrow[L \to \infty]{p} 0$.

A simple proof of this lemma can be found in Ref. [222]. Combining Theorem 5.6 and Lemma 5.1, we have the following corollary.

Corollary 5.2 Assume that Θ is a compact subset in $\mathbb{R}^d$, and $\theta_0 \in \Theta$ is KLID-identifiable. Then we have $\hat{\theta}_{ML} \xrightarrow[L \to \infty]{p} \theta_0$.

According to Corollary 5.2, the KLID-identifiability is a sufficient condition to guarantee the ML estimator to converge to the true value in probability one. This is not surprising since the ML estimator is in essence a special case of the minimum KLID estimator.

5.1.5.2 Histogram-Based Estimation

The histogram-based estimation is a common nonparametric method for density estimation. Suppose the i.i.d. samples $y_{(1)}^M, \ldots, y_{(L)}^M$ take values in a measurable space $\mathfrak{M}$. Let $\mathscr{P}_L = \{A_{L,1}, A_{L,2}, \ldots, A_{L,m_L}\}$, $L = 1, 2, \ldots, m_L$, be a sequence of partitions of $\mathfrak{M}$, with m_L either finite or infinite, such that the σ-measure $0 < v(A_{L,i}) < \infty$ for each i. Then the standard histogram density estimator with respect to v and $\mathscr{P}_L$ is given by:

$$\hat{f}_{his}(y^M) = \mu_L(A_{L,i})/v(A_{L,i}), \quad \text{if} \quad y^M \in A_{L,i} \tag{5.57}$$

where $\mu_L(A_{L,i})$ is the standard empirical measure of $A_{L,i}$, i.e.,

$$\mu_L(A_{L,i}) = \frac{1}{L}\sum_{i=1}^{L} \mathbb{I}(y_{(i)}^M \in A_{L,i}) \tag{5.58}$$

where $\mathbb{I}(\cdot)$ is the indicator function.

According to Ref. [223], under certain conditions, the density estimator $\hat{f}_{his}$ will converge in reversed order information divergence to the true underlying density f_{θ_0}, and the expected KLID

$$\lim_{L \to \infty} E\{D_{KL}^M(\hat{f}_{his} \| f_{\theta_0})\} = 0 \tag{5.59}$$

Since $D_{KL}^M(\hat{f}_{his} \| f_{\theta_0}) \geq 0$, by Markov's inequality [224], for any $\delta > 0$, we have

$$\Pr\{D_{KL}^M(\hat{f}_{his} \| f_{\theta_0}) \geq \delta\} \leq \frac{E\{D_{KL}^M(\hat{f}_{his} \| f_{\theta_0})\}}{\delta} \tag{5.60}$$

It follows that $\forall\,\delta>0$, $\lim\limits_{L\to\infty}\Pr\{D_{\mathrm{KL}}^M(\hat{f}_{\mathrm{his}}\|f_{\theta_0})\geq\delta\}=0$, and for any $\varepsilon>0$ and $\delta>0$ arbitrarily small, there exists an $N(\varepsilon,\delta)<\infty$ such that for $L>N(\varepsilon,\delta)$,

$$\Pr\{D_{\mathrm{KL}}^M(\hat{f}_{\mathrm{his}}\|f_{\theta_0})>\delta\}<\varepsilon \tag{5.61}$$

Thus we have $D_{\mathrm{KL}}^M(\hat{f}_{\mathrm{his}}\|f_{\theta_0})\xrightarrow[L\to\infty]{P}0$. By Theorem 5.6, the following corollary holds.

Corollary 5.3 Assume that Θ is a compact subset in $\mathbb{R}^d$, and $\theta_0\in\Theta$ is KLID-identifiable. Let $\hat{f}_{\mathrm{his}}$ be the standard histogram density estimator satisfying Eq. (5.59). Then we have $\hat{\theta}_{\mathrm{his}}\xrightarrow[L\to\infty]{P}\theta_0$, where $\hat{\theta}_{\mathrm{his}}=\arg\min_{\theta\in\Theta}D_{\mathrm{KL}}^M(\hat{f}_{\mathrm{his}}\|f_\theta)$.

5.1.5.3 Kernel-Based Estimation

The kernel-based estimation (or kernel density estimation, KDE) is another important nonparametric approach for the density estimation. Given an i.i.d. sample $y_{(1)}^M,\ldots,y_{(L)}^M$, the kernel density estimator is

$$\hat{f}_{\mathrm{Ker}}(y^M)=\frac{1}{L}\sum_{i=1}^L K_h(y^M-y_{(i)}^M)=\frac{1}{Lh^{mM}}\sum_{i=1}^L K\left(\frac{y^M-y_{(i)}^M}{h}\right) \tag{5.62}$$

where K is a kernel function satisfying $K\geq0$ and $\int K=1$, $h>0$ is the kernel width.

For the KDE, the following lemma holds (see chapter 9 in Ref. [98] for details).

Lemma 5.2 Assume that K is a fixed kernel, and the kernel width h depends on L only. If $h\to0$ and $Lh^{mM}\to\infty$ as $L\to\infty$, then $\lim\limits_{L\to\infty}E\{\|\hat{f}_{\mathrm{Ker}}-f_{\theta_0}\|_1\}=0$.

From $\lim\limits_{L\to\infty}E\{\|\hat{f}_{\mathrm{Ker}}-f_{\theta_0}\|_1\}=0$, one cannot derive $D_{\mathrm{KL}}^M(\hat{f}_{\mathrm{Ker}}\|f_{\theta_0})\xrightarrow[L\to\infty]{P}0$. And hence, Theorem 5.6 cannot be applied here. However, if the parameter is estimated by minimizing the total variation (not the KLID), the following theorem holds.

Theorem 5.8 Assume that Θ is a compact subset in $\mathbb{R}^d$, $\theta_0\in\Theta$ is KLID-identifiable, and the kernel width h satisfies the conditions in Lemma 5.2. Then we have $\hat{\theta}_{\mathrm{Ker}}\xrightarrow[L\to\infty]{P}\theta_0$, where $\hat{\theta}_{\mathrm{Ker}}=\arg\min_{\theta\in\Theta}\|\hat{f}_{\mathrm{Ker}}-f_\theta\|_1$.

Proof: As $\lim\limits_{L\to\infty}E\{\|\hat{f}_{\mathrm{Ker}}-f_{\theta_0}\|_1\}=0$, by Markov's inequality, we have $\|\hat{f}_{\mathrm{Ker}}-f_{\theta_0}\|_1\xrightarrow[L\to\infty]{P}0$. Following a similar derivation as for Theorem 5.6, one can easily reach the conclusion.

The KLID and the total variation are both special cases of the family of ϕ-divergence [130]. The ϕ-divergence between the PDFs f_{θ_1} and f_{θ_2} is

$$D_\phi(\theta_1 \| \theta_2) = D_\phi(f_{\theta_1} \| f_{\theta_2}) = \int f_{\theta_2}(x)\phi\left(\frac{f_{\theta_1}(x)}{f_{\theta_2}(x)}\right)dx, \quad \phi \in \Phi^* \tag{5.63}$$

where Φ^* is a class of convex functions. The minimum ϕ-divergence estimator is given by [130]:

$$\hat{\theta}_\phi = \arg\min_{\theta \in \Theta} D_\phi(\hat{f}_L \| f_\theta) \tag{5.64}$$

Below we give a more general result, which includes Theorems 5.6 and 5.8 as special cases.

Theorem 5.9 Assume that Θ is a compact subset in $\mathbb{R}^d$, $\theta_0 \in \Theta$ is KLID-identifiable, and for a given $\phi \in \Phi^*$, $\forall\, \theta \in \Theta$, $D_\phi(\hat{f}_L \| f_\theta) \geq \kappa\left(\|\hat{f}_L - f_\theta\|_1\right)$, where function $\kappa(\cdot)$ is strictly increasing over the interval $[0, \infty)$, and $\kappa(0) = 0$. Then, if the density estimate sequence $\{\hat{f}_L\}_{L=1}^\infty$ satisfies $D_\phi(\hat{f}_L \| f_\theta) \xrightarrow[L \to \infty]{P} 0$, we have $\hat{\theta}_\phi \xrightarrow[L \to \infty]{P} \theta_0$, where $\hat{\theta}_\phi$ is the minimum ϕ-divergence estimator.

Proof: Similar to the proof of Theorem 5.6 (omitted).

5.2 Minimum Information Divergence Identification with Reference PDF

Information divergences have been suggested by many authors for the solution of the related problems of system identification. The ML criterion and its extensions (e.g., AIC) can be derived from the KL divergence approach. The information divergence approach is a natural generalization of the LS view. Actually one can think of a "distance" between the actual (empirical) and model distributions of the data, without necessarily introducing the conceptually more demanding concepts of likelihood or posterior. In the following, we introduce a novel system identification approach based on the minimum information divergence criterion.

Apart from conventional methods, the new approach adopts the idea of *PDF shaping* and uses the divergence between the actual error PDF and a reference (or target) PDF (usually with zero mean and a narrow range) as the identification criterion. As illustrated in Figure 5.1, in this scheme, the model parameters are adjusted such that the error distribution tends to the reference distribution. With KLID, the optimal parameters (or weights) of the model can be expressed as:

$$W^* = \arg\min_{W \in \Omega_W} D_{\mathrm{KL}}(p_e \| p_r) = \arg\min_{W \in \Omega_W} \int_{-\infty}^{\infty} p_e(\xi)\log\frac{p_e(\xi)}{p_r(\xi)}d\xi \tag{5.65}$$

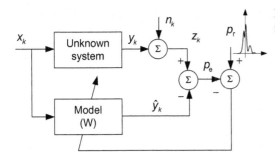

Figure 5.1 Scheme of system identification with a reference PDF.

where p_e and p_r denote, respectively, the actual error PDF and the reference PDF. Other information divergence measures such as ϕ-divergence can also be used but are not considered here.

The above method shapes the error distribution, and can be used to achieve the desired variance or entropy of the error, provided the desired PDF of the error can be achieved. This is expected to be useful in complex signal processing and learning systems. If we choose the δ function as the reference PDF, the identification error will be forced to concentrate around the zero with a sharper peak. This coincides with commonsense predictions about system identification.

It is worth noting that the PDF shaping approaches can be found in other contexts. In the control literature, Karny et al. [225,226] proposed an alternative formulation of stochastic control design problem: the joint distributions of closed-loop variables should be forced to be as close as possible to their desired distributions. This formulation is called the *fully probabilistic control*. Wang et al. [227−229] designed new algorithms to control the shape of the output PDF of a stochastic dynamic system. In adaptive signal processing literature, Sala-Alvarez et al. [230] proposed a general criterion for the design of adaptive systems in digital communications, called the *statistical reference criterion*, which imposes a given PDF at the output of an adaptive system.

It is important to remark that the minimum value of the KLID in Eq. (5.65) may not be zero. In fact, all the possible PDFs of the error are, in general, restricted to a certain set of functions $\mathscr{P}_e$. If the reference PDF is not contained in the possible PDF set, i.e., $p_r \notin \mathscr{P}_e$, we have

$$\min_{W \in \Omega_W} D_{\mathrm{KL}}(p_e\|p_r) = \min_{p_e \in \mathscr{P}_e} D_{\mathrm{KL}}(p_e\|p_r) \neq 0 \qquad (5.66)$$

In this case, the optimal error PDF $p_e^* \triangleq \arg \min_{p_e \in \mathscr{P}_e} D_{\mathrm{KL}}(p_e\|p_r) \neq p_r$, and the reference distribution can never be realized. This is however not a problem of great concern, since our goal is just to make the error distribution closer (not necessarily identical) to the reference distribution.

In some special situations, this new identification method is equivalent to the ML identification. Suppose that in Figure 5.1 the noise n_k is independent of the input x_k, and the unknown system can be exactly identified, i.e., the intrinsic error

$(\bar{e}_k = y_k - \hat{y}_k)$ between the unknown system and the model can be zero. In addition, we assume that the noise PDF p_n is known. In this case, if setting $p_r = p_n$, we have

$$
\begin{aligned}
W^* &= \underset{W \in \Omega_W}{\arg\min}\, D_{KL}(p_e \| p_n) \\
&\overset{(a)}{=} \underset{W \in \Omega_W}{\arg\min}\{D_{KL}(p_e \| p_n) + H(p_e)\} \\
&= \underset{W \in \Omega_W}{\arg\min}\left\{ \int_{-\infty}^{\infty} p_e(\xi)\log\left(\frac{p_e(\xi)}{p_n(\xi)}\right)d\xi - \int_{-\infty}^{\infty} p_e(\xi)\log p_e(\xi)d\xi \right\} \\
&= \underset{W \in \Omega_W}{\arg\min} - \int_{-\infty}^{\infty} p_e(\xi)\log p_n(\xi)d\xi \\
&= \underset{W \in \Omega_W}{\arg\min}\, E_e[-\log p_n(e)] \\
&\approx \underset{W \in \Omega_W}{\arg\max} \sum_{k=1}^{N} \log p_n(e_k) \\
&= \underset{W \in \Omega_W}{\arg\max}\, \log L(W)
\end{aligned}
\tag{5.67}
$$

where (a) comes from the fact that the weight vector minimizing the KLID (when $p_e = p_r = p_n$) also minimizes the error entropy, and $L(W) = \prod_{k=1}^{N} p_n(e_k)$ is the likelihood function.

5.2.1 Some Properties

We present in the following some important properties of the *minimum KLID criterion with reference PDF* (called the *KLID criterion* for short).

The KLID criterion is much different from the minimum error entropy (MEE) criterion. The MEE criterion does not consider the mean of the error due to its invariance to translation. Under MEE criterion, the estimator makes the error PDF as sharp as possible, and neglects the PDF's location. Under KLID criterion, however, the estimator makes the actual error PDF and reference PDF as close as possible (in both shape and location).

The KLID criterion is sensitive to the error mean. This can be easily verified: if p_e and p_r are both Gaussian PDFs with zero mean and unit variance, we have $D_{KL}(p_e \| p_r) = 0$; while if the error mean becomes nonzero, $E(e) = \mu \neq 0$, we have $D_{KL}(p_e \| p_r) = \mu^2/2 \neq 0$. The following theorem suggests that, under certain conditions the mean value of the optimal error PDF under KLID criterion is equal to the mean value of the reference PDF.

Theorem 5.10 Assume that p_e and p_r satisfy:

1. $\forall p_e \in \mathscr{P}_e,\ p_e(e+c) \in \mathscr{P}_e\ (\forall c \in \mathbb{R})$;
2. $\forall p_e \in \mathscr{P}_e,\ q(e) = p_e(e+\mu)$ is an even function, where μ is the mean value of p_e;

3. $q_r(e) = p_r(e + \mu_r)$ is an even and strictly log-concave function, where μ_r is the mean value of p_r.

Then, the mean value of the optimal error PDF (p_e^*) is $\mu_* = \mu_r$.

Proof: Using reduction to absurdity. Suppose $\mu_* \neq \mu_r$, and let $\delta_\mu = \mu_* - \mu_r \neq 0$. Denote $q^*(e) = p_e^*(e + \mu_*)$ and $q_\delta^*(e) = p_e^*(e + \delta_\mu)$. According to the assumptions, $q^*, q_\delta^* \in \mathscr{P}_e$, and $q^*(e)$ is an even function. Then we have

$$D_{KL}(q_\delta^*(e)\|p_r(e)) = D_{KL}(p_e^*(e + \delta_\mu)\|p_r(e))$$

$$\overset{(a)}{=} D_{KL}(p_e^*(e + \mu_r + \delta_\mu)\|p_r(e + \mu_r))$$

$$= D_{KL}(p_e^*(e + \mu_*)\|p_r(e + \mu_r)) = D_{KL}(q^*(e)\|q_r(e))$$

$$= -H(q^*) - \int_R q^*(e)\log q_r(e)de$$

$$\overset{(b)}{<} -H(q^*) - \int_R q^*(e)\log\sqrt{q_r(e + \delta_\mu)q_r(e - \delta_\mu)}de$$

$$= -H(q^*) - \frac{1}{2}\int_R q^*(e)\log q_r(e + \delta_\mu)de - \frac{1}{2}\int_R q^*(e)\log q_r(e - \delta_\mu)de$$

$$\overset{(c)}{=} -H(q^*) - \frac{1}{2}\int_R q^*(e)\log q_r(e + \delta_\mu)de - \frac{1}{2}\int_R q^*(e)\log q_r(e + \delta_\mu)de$$

$$= -H(q^*) - \int_R q^*(e)\log q_r(e + \delta_\mu)de = D_{KL}(q^*(e)\|q_r(e + \delta_\mu))$$

$$\overset{(d)}{=} D_{KL}(q^*(e - \delta_\mu)\|q_r(e)) = D_{KL}(p_e^*(e + \mu_r)\|p_r(e + \mu_r))$$

$$\overset{(e)}{=} D_{KL}(p_e^*(e)\|p_r(e)) \tag{5.68}$$

where (a), (d), and (e) follow from the *shift-invariance* of the KLID, (b) is because $q_r(e)$ is strictly log-concave, and (c) is because $q^*(e)$ and $q_r(e)$ are even functions. Therefore, $\exists q_\delta^* \in \mathscr{P}_e$, such that

$$D_{KL}(q_\delta^*\|p_r) < D_{KL}(p_e^*\|p_r) \tag{5.69}$$

This contradicts with $D_{KL}(p_e^*\|p_r) = \min_{p_e \in \mathscr{P}_e} D_{KL}(p_e\|p_r)$. And hence, $\mu_* = \mu_r$ holds.

On the other hand, the KLID criterion is also closely related to the MEE criterion. The next theorem provides an upper bound on the error entropy under the constraint that the KLID is bounded.

Theorem 5.11 Let the reference PDF p_r be a zero-mean Gaussian PDF with variance σ^2. If the error PDF p_e satisfies

$$D_{KL}(p_e \| p_r) \leq c \tag{5.70}$$

where $c > 0$ is a positive constant, then the error entropy $H(e)$ satisfies

$$H(e) \leq H(p_r) + \log \sqrt{(1 + \lambda)/\lambda} \tag{5.71}$$

where $\lambda > 0$ is the solution of the following equation:

$$\log \left(\frac{\lambda}{1 + \lambda} \right) + \frac{1}{\lambda} = 2c \tag{5.72}$$

Proof: Denote $\mathcal{B}(p_r, c)$ the collection of all the error PDFs that satisfy $D_{KL}(p_e \| p_r) \leq c$. Clearly, this is a convex set. $\forall p_e \in \mathcal{B}(p_r, c)$, we have

$$H(e) \leq \max_{p \in \mathcal{B}(p_r, c)} H(p) = \max_{p \in \mathcal{B}(p_r, c)} \left\{ -\int_{-\infty}^{+\infty} p(x) \log p(x) dx \right\} \tag{5.73}$$

In order to solve the error distribution that achieves the maximum entropy, we create the Lagrangian:

$$L(p, \theta, \lambda) = -\int_R p \log p \, dx + \theta(1 - \int_R p \, dx) + \lambda \left(c - \int_R p \log \left(\frac{p}{p_r} \right) dx \right) \tag{5.74}$$

where θ and λ are the Lagrange multipliers. When $\lambda > 0$, $L(p, \theta, \lambda)$ is a concave function of $p \in \mathcal{B}(p_r, c)$. If β is a function such that $p + \varepsilon\beta \in \mathcal{B}(p_r, c)$ for ε sufficiently small, the Gateaux derivative of L with respect to p is given by:

$$\lim_{\varepsilon \to 0} \frac{1}{\varepsilon} \left\{ L(p + \varepsilon\beta, \theta, \lambda) - L(p, \theta, \lambda) \right\} \tag{5.75}$$
$$= \int_R \{ -\theta - (1 + \lambda) - (1 + \lambda)\log p + \lambda \log p_r \}\beta \, dx$$

If it is zero for all β, we have

$$\log p = -1 - \frac{\theta}{1 + \lambda} + \frac{\lambda}{1 + \lambda} \log p_r \tag{5.76}$$

Thus, if $\lambda > 0$ (such that L is a concave function of p), the error PDF that achicves the maximum entropy exists and is given by

$$p_0 = \exp \left(-1 - \frac{\theta}{1 + \lambda} \right) \exp \left(\frac{\lambda}{1 + \lambda} \log p_r \right) \tag{5.77}$$

According to the assumptions, $p_r(x) = (1/\sqrt{2\pi}\sigma)\exp(-x^2/2\sigma^2)$. It follows that

$$p_0 = \exp\left(-1 - \frac{\theta}{1+\lambda}\right) \times \left(\frac{1}{\sqrt{2\pi}\sigma}\right)^{\frac{\lambda}{1+\lambda}} \times \exp\left(-\frac{x^2}{2\sigma_0^2}\right) \qquad (5.78)$$

where $\sigma_0^2 = (1 + \lambda/\lambda)\sigma^2$. Obviously, p_0 is a Gaussian density, and we have

$$\exp\left(-1 - \frac{\theta}{1+\lambda}\right) \times \left(\frac{1}{\sqrt{2\pi}\sigma}\right)^{\frac{\lambda}{1+\lambda}} = \frac{1}{\sqrt{2\pi}\sigma_0} \qquad (5.79)$$

So θ can be determined as

$$\theta = -(1 + \lambda) - (1 + \lambda) \times \log\left\{\sqrt{\frac{\lambda}{1+\lambda}} \times \left(\sqrt{2\pi}\sigma\right)^{\frac{-1}{1+\lambda}}\right\} \qquad (5.80)$$

In order to determine the value of λ, we use the Kuhn–Tucker condition:

$$\lambda(c - D_{KL}(p_0\|p_r)) = 0 \qquad (5.81)$$

When $\lambda > 0$, we have $D_{KL}(p_0\|p_r) = c$, that is,

$$D_{KL}(p_0\|p_r) = \int_{\mathbb{R}} p_0(x)\log\left(\frac{p_0(x)}{p_r(x)}\right)dx = \frac{1}{2}\left\{\log\left(\frac{\lambda}{1+\lambda}\right) + \frac{1}{\lambda}\right\} = c \qquad (5.82)$$

Therefore, λ is the solution of the Eq. (5.72).

Define the function $\varphi(\lambda) = \log(\lambda/(1+\lambda)) + (1/\lambda)$. It is easy to verify that $\varphi(\lambda)$ is continuous and monotonically decreasing over interval $(0, +\infty)$. Since $\lim_{\lambda \to 0+} \varphi(\lambda) = +\infty$, $\lim_{\lambda \to +\infty} \varphi(\lambda) = 0$, and $c > 0$, the equation $\varphi(\lambda) = 2c$ certainly has a solution in $(0, +\infty)$.

From the previous derivations, one may easily obtain:

$$H(e) \leq \max_{p \in \mathcal{B}(p_r, c)} H(p) = H(p_0) = H(p_r) + \log\sqrt{(1+\lambda)/\lambda} \qquad (5.83)$$

The above theorem indicates that, under the KLID constraint $D_{KL}(p_e\|p_r) \leq c$, the error entropy is upper bounded by the reference entropy plus a certain constant. In particular, when $c \to 0+$, we have $\lambda \to +\infty$ and $\max_{p \in \mathcal{B}(p_r, c)} H(p) \to H(p_r)$. Therefore, if one chooses a reference PDF with small entropy, the error entropy will also be confined within small values. In practice, the reference PDF is, in general, chosen as a PDF with zero mean and small entropy (e.g., the δ distribution at zero).

In most practical situations, the error PDF is unknown and needs to be estimated from samples. There is always a bias in the density estimation; in order to offset the influence of the bias, one can use the same method to estimate the reference density based on the samples drawn from the reference PDF. Let $S_e = \{e_1 \quad e_2 \quad \cdots \quad e_N\}$ and $S_r = \{e_1^{(r)} \quad e_2^{(r)} \quad \cdots \quad e_N^{(r)}\}$ be respectively the actual and reference error samples. The KDEs of p_e and p_r will be

$$
\begin{cases}
\hat{p}_e(e) = \dfrac{1}{N} \sum_{e_k \in S_e} K_{h_e}(e - e_k) \\[3mm]
\hat{p}_r(e) = \dfrac{1}{N} \sum_{e_k^{(r)} \in S_r} K_{h_r}(e - e_k^{(r)})
\end{cases}
\tag{5.84}
$$

where h_e and h_r are corresponding kernel widths. Using the estimated PDFs, one may obtain the empirical KLID criterion $\hat{D}_{KL}(p_e \| p_r) = D_{KL}(\hat{p}_e \| \hat{p}_r)$.

Theorem 5.12 The empirical KLID $\hat{D}_{KL}(p_e \| p_r) \geq 0$, with equality if and only if $\hat{p}_e = \hat{p}_r$.

Proof: $\forall x > 0$, we have $\log x \geq 1 - \dfrac{1}{x}$, with equality if and only if $x = 1$. And hence

$$
\hat{D}_{KL}(p_e \| p_r) = \int \hat{p}_e(e) \log \frac{\hat{p}_e(e)}{\hat{p}_r(e)} \, de
$$

$$
\geq \int \hat{p}_e(e) \left(1 - \frac{\hat{p}_r(e)}{\hat{p}_e(e)} \right) de
\tag{5.85}
$$

$$
= \int (\hat{p}_e(e) - \hat{p}_r(e)) de = 0
$$

with equality if and only if $\hat{p}_e = \hat{p}_r$.

Lemma 5.3 If two PDFs $p_1(x)$ and $p_2(x)$ are bounded, then [231]:

$$
D_{KL}(p_1 \| p_2) \geq \frac{1}{\alpha} \int_{-\infty}^{\infty} (p_1(x) - p_2(x))^2 \, dx
\tag{5.86}
$$

where $\alpha = 2 \max\{\sup p_1(x), \sup p_2(x)\}$.

Theorem 5.13 If $K_h(\cdot)$ is a Gaussian kernel function, $K_h(x) = \dfrac{1}{\sqrt{2\pi}h}\exp\left\{\dfrac{-x^2}{2h^2}\right\}$, then

$$\hat{D}_{\mathrm{KL}}(p_e\|p_r) \geq \frac{\sqrt{2\pi}\{\min(h_e, h_r)\}}{2}\int_{-\infty}^{\infty}(\hat{p}_e(e) - \hat{p}_r(e))^2\mathrm{d}e \tag{5.87}$$

Proof: Since Gaussian kernel is bounded, and $\sup K_h(x) = (1/\sqrt{2\pi}h)$, the kernel-based density estimates $\hat{p}_e$ and $\hat{p}_r$ will also be bounded, and

$$\begin{cases} \sup \hat{p}_e(e) = \sup_e\left\{\dfrac{1}{N}\displaystyle\sum_{e_k \in S_e} K_{h_e}(e - e_k)\right\} \leq \dfrac{1}{\sqrt{2\pi}h_e} \\[4mm] \sup \hat{p}_r(e) = \sup_e\left\{\dfrac{1}{N}\displaystyle\sum_{e_k^{(r)} \in S_r} K_{h_r}(e - e_k^{(r)})\right\} \leq \dfrac{1}{\sqrt{2\pi}h_r} \end{cases} \tag{5.88}$$

By Lemma 5.3, we have

$$\hat{D}_{\mathrm{KL}}(p_e\|p_r) \geq \frac{1}{\alpha}\int_{-\infty}^{\infty}(\hat{p}_e(e) - \hat{p}_r(e))^2\,\mathrm{d}e \tag{5.89}$$

where

$$\begin{aligned} \alpha &= 2\max\{\sup \hat{p}_e, \sup \hat{p}_r\} \\[2mm] &\leq 2\max\left\{\frac{1}{\sqrt{2\pi}h_e}, \frac{1}{\sqrt{2\pi}h_r}\right\} \\[2mm] &= \frac{2}{\sqrt{2\pi}\{\min(h_e, h_r)\}} \end{aligned} \tag{5.90}$$

Then we obtain Eq. (5.87).

The above theorem suggests that convergence in KLID ensures the convergence in L_2 distance ($\sqrt{\int_{-\infty}^{\infty}(p(x) - q(x))^2\mathrm{d}x}$).

Before giving Theorem 5.14, we introduce some notations. By rearranging the samples in S_e and S_r, one obtains the increasing sequences:

$$\begin{cases} S'_e = \{e_{k_1} \quad e_{k_2} \quad \cdots \quad e_{k_N}\} \\ S'_r = \{e_{k_1}^{(r)} \quad e_{k_2}^{(r)} \quad \cdots \quad e_{k_N}^{(r)}\} \end{cases} \tag{5.91}$$

where $e_{k_1} = \cdots = e_{k_{p_1}} < e_{k_{p_1}+1} = \cdots = e_{k_{p_2}} < \cdots = e_{k_N}$, $e_{k_1}^{(r)} = \cdots = e_{k_{q_1}}^{(r)} < e_{k_{q_1}+1}^{(r)} = \cdots$ $= e_{k_{q_2}}^{(r)} < \cdots = e_{k_N}^{(r)}$. Denote $\widehat{S}_e = \{e_{k_{p_1}} \quad e_{k_{p_2}} \quad \cdots \quad e_{k_N}\}$, and $\widehat{S}_r = \{e_{k_{q_1}}^{(r)} \quad e_{k_{q_2}}^{(r)}$ $\cdots e_{k_N}^{(r)}\}$.

Theorem 5.14 If K_h is a Gaussian kernel function, $h_e = h_r = h$, then $\hat{D}_{KL}(p_e \| p_r) = 0$ if and only if $S'_e = S'_r$.

Proof: By Theorem 5.12, it suffices to prove that $\hat{p}_e(e) = \hat{p}_r(e)$ if and only if $S'_e = S'_r$.

Sufficiency: If $S'_e = S'_r$, we have

$$\hat{p}_e(e) = \frac{1}{N} \sum_{e_k \in S_e} K_h(e - e_k)$$

$$= \frac{1}{N} \sum_{e_k^{(r)} \in S_r} K_h(e - e_k^{(r)}) = \hat{p}_r(e) \tag{5.92}$$

Necessity: If $\hat{p}_e(e) = \hat{p}_r(e)$, then we have

$$f(e) \triangleq \sqrt{2\pi}hN \times (\hat{p}_e(e) - \hat{p}_r(e))$$

$$= \sum_{k=1}^{N} \exp\left\{-\frac{1}{2h^2}(e - e_k)^2\right\} - \sum_{k=1}^{N} \exp\left\{-\frac{1}{2h^2}(e - e_k^{(r)})^2\right\} \tag{5.93}$$

$$= 0$$

Let $\widehat{S} = \widehat{S}_e \cup \widehat{S}_r = \{\xi_1 \quad \xi_2 \quad \cdots \quad \xi_M\}$, where $\xi_i < \xi_j (\forall i < j), M \leq 2N$. Then,

$$f(e) = \sum_{k=1}^{M} \lambda_k \exp\left\{-\frac{1}{2h^2}(e - \xi_k)^2\right\} = 0 \tag{5.94}$$

where $\lambda_k = |S_e^{(\xi_k)}| - |S_r^{(\xi_k)}|$, $S_e^{(\xi_k)} = \{e_l | e_l \in S'_e, e_l = \xi_k\}$, $S_r^{(\xi_k)} = \{e_l^{(r)} | e_l^{(r)} \in S'_r, e_l^{(r)} = \xi_k\}$.
Since $\forall e \in \mathbb{R}, f(e) = 0$, we have $f(\xi_i) = 0$, $i = 1, 2, \ldots, M$. It follows that

$$
\begin{bmatrix}
1 & \exp\left\{-\dfrac{(\xi_1-\xi_2)^2}{2h^2}\right\} & \cdots & \exp\left\{-\dfrac{(\xi_1-\xi_M)^2}{2h^2}\right\} \\
\exp\left\{-\dfrac{(\xi_2-\xi_1)^2}{2h^2}\right\} & 1 & \cdots & \exp\left\{-\dfrac{(\xi_2-\xi_M)^2}{2h^2}\right\} \\
\vdots & \vdots & \ddots & \vdots \\
\exp\left\{-\dfrac{(\xi_M-\xi_1)^2}{2h^2}\right\} & \exp\left\{-\dfrac{(\xi_M-\xi_2)^2}{2h^2}\right\} & \cdots & 1
\end{bmatrix}
\begin{bmatrix}
\lambda_1 \\ \lambda_2 \\ \vdots \\ \lambda_M
\end{bmatrix}
= \Phi\vec{\lambda} = 0
$$

$$(5.95)$$

As Φ is a symmetric and positive definite matrix ($\det\Phi \neq 0$), we get $\vec{\lambda} = \mathbf{0}$, that is

$$\lambda_k = \left|S_e^{(\xi_k)}\right| - \left|S_r^{(\xi_k)}\right| = 0, \quad k = 1, 2, \ldots, M \tag{5.96}$$

Thus, $\left|S_e^{(\xi_k)}\right| = \left|S_r^{(\xi_k)}\right|$, and

$$
S'_e = \left\{ \underbrace{\xi_1, \ldots, \xi_1}_{\left|S_e^{(\xi_1)}\right|}, \underbrace{\xi_2, \ldots, \xi_2}_{\left|S_e^{(\xi_2)}\right|} \cdots \underbrace{\xi_M, \ldots, \xi_M}_{\left|S_e^{(\xi_M)}\right|} \right\}
$$

$$(5.97)$$

$$
= \left\{ \underbrace{\xi_1, \ldots, \xi_1}_{\left|S_r^{(\xi_M)}\right|}, \underbrace{\xi_2, \ldots, \xi_2}_{\left|S_r^{(\xi_2)}\right|} \cdots \underbrace{\xi_M, \ldots, \xi_M}_{\left|S_r^{(\xi_M)}\right|} \right\} = S'_r
$$

This completes the proof.

Theorem 5.14 indicates that, under certain conditions the zero value of the empirical KLID occurs only when the actual and reference sample sets are identical.

Based on the sample sets S'_e and S'_r, one can calculate the empirical distribution:

$$F_{S'_e}(e) = \frac{n_e}{N}, \quad F_{S'_r}(e) = \frac{n_e^{(r)}}{N} \tag{5.98}$$

where $n_e = \left|\{e_k | e_k \in S'_e, e_k \le e\}\right|$, and $n_e^{(r)} = \left|\{e_k^{(r)} | e_k^{(r)} \in S'_r, e_k^{(r)} \le e\}\right|$. According to the limit theorem in probability theory [224], we have

$$
\begin{cases}
F_{S'_e}(e) \to F_e(e) = \int_{-\infty}^{e} p_e(\tau)d\tau \\
F_{S'_r}(e) \to F_r(e) = \int_{-\infty}^{e} p_r(\tau)d\tau
\end{cases}
\quad \text{as} \quad N \to \infty
\tag{5.99}
$$

If $S'_e = S'_r$, and N is large enough, we have $\int_{-\infty}^{e} p_e(\tau)d\tau \approx F_{S'_e}(e) = F_{S'_r}(e) \approx \int_{-\infty}^{e} p_r(\tau)d\tau$, and hence $p_e(e) \approx p_r(e)$. Therefore, when the empirical KLID approaches zero, the actual error PDF will be approximately identical with the reference PDF.

5.2.2 Identification Algorithm

In the following, we derive a stochastic gradient−based identification algorithm under the minimum KLID criterion with a reference PDF. Since the KLID is not symmetric, we use the symmetric version of KLID (also referred to as the J-information divergence):

$$
\begin{aligned}
J(p_e \| p_r) &= D_{KL}(p_e \| p_r) + D_{KL}(p_r \| p_e) \\
&= E_e\left[\log\left(\frac{p_e(e)}{p_r(e)}\right)\right] + E_r\left[\log\left(\frac{p_r(e)}{p_e(e)}\right)\right]
\end{aligned}
\tag{5.100}
$$

By dropping off the expectation operators E_e and E_r, and plugging in the estimated PDFs, one may obtain the estimated instantaneous value of J-information divergence:

$$
\hat{J}_k(p_e \| p_r) = \log\frac{\hat{p}_e(e_k)}{\hat{p}_r(e_k)} + \log\frac{\hat{p}_r(e_k^{(r)})}{\hat{p}_e(e_k^{(r)})}
\tag{5.101}
$$

where $\hat{p}_e(\cdot)$ and $\hat{p}_r(\cdot)$ are

$$
\begin{cases}
\hat{p}_e(\xi) = \dfrac{1}{L} \displaystyle\sum_{i=k-L+1}^{k} K_{h_e}(\xi - e_i) \\
\hat{p}_r(\xi) = \dfrac{1}{L} \displaystyle\sum_{i=k-L+1}^{k} K_{h_r}(\xi - e_i^{(r)})
\end{cases}
\tag{5.102}
$$

Then a stochastic gradient−based algorithm can be readily derived as follows:

$$W_{k+1} = W_k - \eta \frac{\partial}{\partial W} \hat{J}_k(p_e \| p_r)$$

$$= W_k - \eta \left\{ \frac{\partial}{\partial W} \log \frac{\hat{p}_e(e_k)}{\hat{p}_r(e_k)} + \frac{\partial}{\partial W} \log \frac{\hat{p}_r(e_k^{(r)})}{\hat{p}_e(e_k^{(r)})} \right\}$$

$$= W_k - \eta \left\{ \frac{\frac{\partial}{\partial W} \hat{p}_e(e_k)}{\hat{p}_e(e_k)} - \frac{\frac{\partial}{\partial W} \hat{p}_r(e_k)}{\hat{p}_r(e_k)} - \frac{\frac{\partial}{\partial W} \hat{p}_e(e_k^{(r)})}{\hat{p}_e(e_k^{(r)})} \right\}$$

(5.103)

where

$$\begin{cases} \dfrac{\partial}{\partial W} \hat{p}_e(e_k) = -\dfrac{1}{L} \sum_{i=k-L+1}^{k} \left\{ K'_{h_e}(e_k - e_i) \left(\dfrac{\partial}{\partial W} \hat{y}_k - \dfrac{\partial}{\partial W} \hat{y}_i \right) \right\} \\[3mm] \dfrac{\partial}{\partial W} \hat{p}_r(e_k) = -\dfrac{1}{L} \left\{ \sum_{i=k-L+1}^{k} K'_{h_r}(e_k - e_i^{(r)}) \right\} \left(\dfrac{\partial}{\partial W} \hat{y}_k \right) \\[3mm] \dfrac{\partial}{\partial W} \hat{p}_e(e_k^{(r)}) = \dfrac{1}{L} \sum_{i=k-L+1}^{k} \left\{ K'_{h_e}(e_k^{(r)} - e_i) \left(\dfrac{\partial}{\partial W} \hat{y}_i \right) \right\} \end{cases}$$

(5.104)

This algorithm is called the stochastic information divergence gradient (SIDG) algorithm [125,126].

In order to achieve an error distribution with zero mean and small entropy, one can choose the δ function at zero as the reference PDF. It is, however, worth noting that the δ function is not always the best choice. In many situations, the desired error distribution may be far from the δ distribution. In practice, the desired error distribution can be estimated from some prior knowledge or preliminary identification results.

Remark: Strictly speaking, if one selects the δ function as the reference distribution, the information divergence will be undefined (ill-posed). In practical applications, however, we often use the estimated information divergence as an alternative cost function, where the actual and reference error distributions are both estimated by KDE approach (usually with the same kernel width). It is easy to verify that, for the δ distribution, the estimated PDF is actually the kernel function. In this case, the estimated divergence will always be valid.

5.2.3 Simulation Examples

Example 5.3 Consider the FIR system identification [126]:

$$\begin{cases} z_k = w_0^* x_k + w_1^* x_{k-1} \\ \hat{y}_k = w_0 x_k + w_1 x_{k-1} \end{cases} \tag{5.105}$$

where the true weight vector $W^* = [w_0^*, w_1^*]^T = [1.0, 0.5]^T$. The input signal $\{x_k\}$ is assumed to be a zero-mean white Gaussian process with unit power.

We show that the optimal solution under information divergence criterion may be not unique. Suppose the reference PDF p_r is Gaussian PDF with zero mean and variance ε^2. The J-information divergence between p_e and p_r can be calculated as:

$$J(p_e \| p_r) = \frac{1}{2} \times \frac{[(w_0 - 1)^2 + (w_1 - 0.5)^2 - \varepsilon^2]^2}{\varepsilon^2 \times [(w_0 - 1)^2 + (w_1 - 0.5)^2]} \tag{5.106}$$

Clearly, there are infinitely many weight pairs (w_0, w_1) that satisfy $J(p_e \| p_r) = 0$. In fact, any weight pair (w_0, w_1) that lies on the circle $(w_0 - 1)^2 + (w_1 - 0.5)^2 = \varepsilon^2$ will be an optimal solution. In this case, the system parameters are not identifiable. However, when $\varepsilon \to 0+$, the circle will shrink to a point and all the solutions will converge to a unique solution $(1.0, 0.5)$. For the case $\varepsilon = 0.5$, the 3D surface of the J-information divergence is depicted in Figure 5.2. Figure 5.3 draws the convergence trajectories of the weight pair (w_0, w_1) learned by the SIDG algorithm,

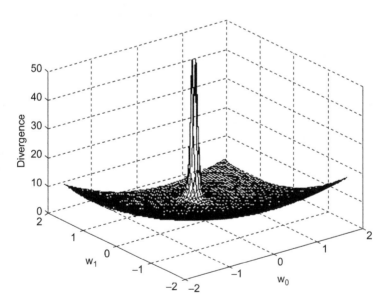

Figure 5.2 3D surface of J-information divergence.
Source: Adapted from Ref. [126].

starting from the initial point $(2.0, -2.0)$. As expected, these trajectories converge to the circles centered at $(1.0, 0.5)$. When $\varepsilon = 0.01$, the weight pair (w_0, w_1) will converge to $(1.0047, 0.4888)$, which is very close to the true weight vector.

Example 5.4 Identification of the hybrid system (switch system) [125]:

$$
\begin{cases}
x_{k+1} = \begin{cases} 2x_k + u_k + r_k, & \text{if } x_k \leq 0 \\ -1.5x_k + u_k + r_k, & \text{if } x_k > 0 \end{cases} \\
y_k = x_k + m_k
\end{cases}
\tag{5.107}
$$

where x_k is the state variable, u_k is the input, r_k is the process noise, and m_k is the measurement noise. This system can be written in a parameterized form (r_k merging into u_k) [125]:

$$
b_{\lambda_k}^T x + a_{\lambda_k}^T m = 0
\tag{5.108}
$$

where $\lambda_k = 1, 2$ is the mode index, $x = [u_{k-1}, y_{k-1}, -y_k]^T$, $m = [m_{k-1}, m_k]^T$, $b_i = [c_{1,i}, a_{1,i}, 1]^T$, and $a_i = [-a_{1,i}, 1]^T$. In this example, $b_1 = [1, 2, 1]^T$ and $b_2 = [1, -1.5, 1]^T$. Based on the parameterized form (Eq. (5.108)), one can establish the *noisy hybrid decoupling polynomial* (NHDP) [125]. By expanding the NHDP and ignoring the higher-order components of the noise, we obtain the first-order approximation (FOA) model. In Ref. [125], the SIDG algorithm (based on the FOA model) was applied to identify the above hybrid system. Figure 5.4 shows the

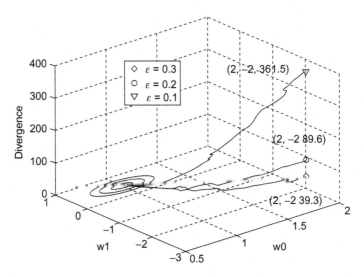

Figure 5.3 Convergence trajectories of weight pair (w_0, w_1).
Source: Adapted from Ref. [126].

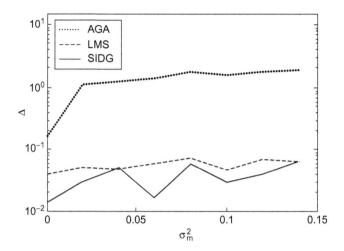

Figure 5.4 Identification performance for different measurement noise powers.
Source: Adapted from Ref. [125].

identification performance for different measurement noise powers σ_m^2. For comparison purpose, we also draw the performance of the least mean square (LMS) and algebraic geometric approach [232]. In Figure 5.4, the identification performance Δ is defined as:

$$\Delta = \max_{i=1,2} \left(\min_{j=1,2} \frac{\|\hat{\boldsymbol{b}}_i - \boldsymbol{b}_j\|}{\|\boldsymbol{b}_j\|} \right) \tag{5.109}$$

In the simulation, the δ function is selected as the reference PDF for SIDG algorithm. Simulation results indicate that the SIDG algorithm can achieve a better performance.

To further verify the performance of the SIDG algorithm, we consider the case in which $r_k = 0$ and m_k is uniformly distributed in the range of $[-1, -0.5] \cup [0.5, 1]$. The reference samples are set to $S_r = [-2, -1.6, -1.6, -0.3, -0.3, -0.3, 0.3, 0.3, 2, 2.2]$ according to some preliminary identification results. Figure 5.5 shows the scatter graphs of the estimated parameter vector $\hat{\boldsymbol{b}}_i$ (with 300 simulation runs), where (A) and (B) correspond, respectively, to the LMS and SIDG algorithms. In each graph, there are two clusters. Evidently, the clusters generated by SIDG are more compact than those generated by LMS, and the centers of the former are closer to the true values than those of the latter (the true values are $\{c_1, a_1\}_1 = \{1, 2\}$ and $\{c_1, a_1\}_2 = \{1, -1.5\}$). The

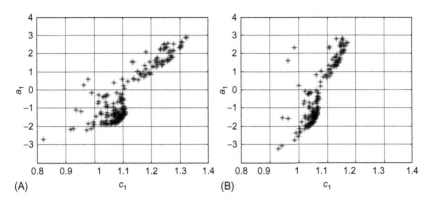

Figure 5.5 Scatter graphs of the estimated parameter vector $\hat{\boldsymbol{b}}_i$: (A) LMS and (B) SIDG.

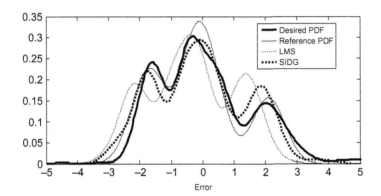

Figure 5.6 Comparison of the error PDFs.
Source: Adapted from Ref. [125].

involved error PDFs are illustrated in Figure 5.6. As one can see, the error distribution produced by SIDG is closer to the desired error distribution.

5.2.4 Adaptive Infinite Impulsive Response Filter with Euclidean Distance Criterion

In Ref. [233], the Euclidean distance criterion (EDC), which can be regarded as a special case of the information divergence criterion with a reference PDF, was successfully applied to develop the global optimization algorithms for adaptive infinite impulsive response (IIR) filters. In the following, we give a brief introduction of this approach.

The EDC for the adaptive IIR filters is defined as the Euclidean distance (or L_2 distance) between the error PDF and δ function [233]:

$$\text{EDC} = \int_{-\infty}^{\infty} (p_e(\xi) - \delta(\xi))^2 d\xi \tag{5.110}$$

The above formula can be expanded as:

$$\text{EDC} = \int_{-\infty}^{\infty} p_e^2(\xi)d\xi - 2p_e(0) + c \tag{5.111}$$

where c stands for the parts of this Euclidean distance measure that do not depend on the error distribution. By dropping c, the EDC can be simplified to

$$\text{EDC} = V_2(e) - 2p_e(0) \tag{5.112}$$

where $V_2(e) = \int_{-\infty}^{\infty} p_e^2(\xi)d\xi$ is the quadratic information potential of the error.

By substituting the kernel density estimator (usually with Gaussian kernel G_h) for the error PDF in the integral, one may obtain the empirical EDC:

$$\widehat{EDC} = \int_{-\infty}^{\infty} \hat{p}_e^2(\xi)d\xi - 2\hat{p}_e(0)$$

$$= \int_{-\infty}^{\infty} \left(\frac{1}{L} \sum_{i=k-L+1}^{k} G_h(\xi - e_i)\right)^2 d\xi - \frac{2}{L} \sum_{i=k-L+1}^{k} G_h(0 - e_i) \tag{5.113}$$

$$= \frac{1}{L^2} \sum_{j=k-L+1}^{k} \sum_{i=k-L+1}^{k} G_{\sqrt{2}h}(e_i - e_j) - \frac{2}{L} \sum_{i=k-L+1}^{k} G_h(e_i)$$

A gradient-based identification algorithm can then be derived as follows:

$$W_{k+1} = W_k - \eta \frac{\partial}{\partial W} \widehat{EDC}$$

$$= W_k - \frac{\eta}{2L^2 h^2} \sum_{j=k-L+1}^{k} \sum_{i=k-L+1}^{k} \left\{ (e_i - e_j) G_{\sqrt{2}h}(e_i - e_j) \left(\frac{\partial \hat{y}_i}{\partial W} - \frac{\partial \hat{y}_j}{\partial W}\right) \right\}$$

$$+ \frac{2\eta}{Lh^2} \sum_{i=k-L+1}^{k} \left\{ e_i G_h(e_i) \frac{\partial \hat{y}_i}{\partial W} \right\} \tag{5.114}$$

where the gradient $(\partial \hat{y}_i / \partial W)$ depends on the model structure. Below we derive this gradient for the IIR filters.

Let us consider the following IIR filter:

$$\hat{y}_k = \sum_{i=0}^{n_b} b_i x_{k-i} + \sum_{j=1}^{n_a} a_j \hat{y}_{k-j} \tag{5.115}$$

which can be written in the form

$$\hat{y}_k = \varphi_k^T W \tag{5.116}$$

where $\varphi_k = [x_k, \ldots, x_{k-n_b}, \hat{y}_{k-1}, \ldots, \hat{y}_{k-n_a}]^T$, $W = [b_0, \ldots, b_{n_b}, a_1, \ldots, a_{n_a}]^T$. Then we can derive

$$
\begin{aligned}
\partial \hat{y}_k / \partial W &= \partial (W^T \varphi_k) / \partial W \\
&= (\partial W^T / \partial W) \varphi_k + (\partial \varphi_k^T / \partial W) W \\
&= \varphi_k + \sum_{j=1}^{n_a} a_j (\partial \hat{y}_{k-j} / \partial W)
\end{aligned} \tag{5.117}
$$

In Eq. (5.117), the parameter gradient is calculated in a recursive manner.

Example 5.5 Identifying the following unknown system [233]:

$$G^*(z) = \frac{0.05 + 0.4 z^{-1}}{1 - 1.1314 z^{-1} + 0.25 z^{-2}} \tag{5.118}$$

The adaptive model is chosen to be the reduced order IIR filter

$$G(z) = \frac{b}{1 - a z^{-1}} \tag{5.119}$$

The main goal is to determine the values of the coefficients (or weights) $\{a, b\}$, such that the EDC is minimized. Assume that the error is Gaussian distributed, $e_k \sim \mathcal{N}(\mu_e, \sigma_e^2)$. Then, the empirical EDC can be approximately calculated as [233]:

$$\widehat{EDC} \approx \frac{1}{\sqrt{4\pi(\sigma_e^2 + h^2)}} - \frac{2}{\sqrt{2\pi(\sigma_e^2 + h^2)}} \exp\left\{ -\frac{\mu_e^2}{2(\sigma_e^2 + h^2)} \right\} \tag{5.120}$$

where h is the kernel width. Figure 5.7 shows the contours of the EDC performance surface in different h (the input signal is assumed to be a white Gaussian noise

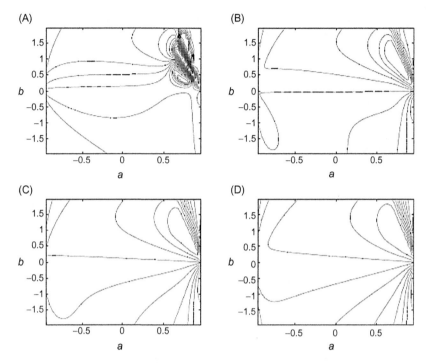

Figure 5.7 Contours of the EDC performance surface: (A) $h^2 = 0$; (B) $h^2 = 1$; (C) $h^2 = 2$; and (D) $h^2 = 3$.
Source: Adapted from Ref. [233].

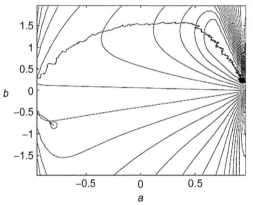

Figure 5.8 Weight convergence trajectory under EDC.
Source: Adapted from Ref. [233].

with zero mean and unit variance). As one can see, the local minima of the performance surface have disappeared with large kernel width. Thus, by carefully controlling the kernel width, the algorithm can converge to the global minimum. The convergence trajectory of the adaptation process with the weight approaching to the global minimum is shown in Figure 5.8.

6 System Identification Based on Mutual Information Criteria

As a central concept in communication theory, mutual information measures the amount of information that one random variable contains about another. The larger the mutual information between two random variables is, the more information they share, and the better the estimation algorithm can be. Typically, there are two mutual information-based identification criteria: the minimum mutual information (MinMI) and the maximum mutual information (MaxMI) criteria. The MinMI criterion tries to minimize the mutual information between the identification error and the input signal such that the error signal contains as little as possible information about the input,[1] while the MaxMI criterion aims to maximize the mutual information between the system output and the model output such that the model contains as much as possible information about the system in their outputs. Although the MinMI criterion is essentially equivalent to the minimum error entropy (MEE) criterion, their physical meanings are different. The MaxMI criterion is somewhat similar to the Infomax principle, an optimization principle for neural networks and other information processing systems. They are, however, different in their concepts. The Infomax states that a function that maps a set of input values I to a set of output values O should be chosen (or learned) so as to maximize the mutual information between I and O, subject to a set of specified constraints and/or noise processes. In the following, we first discuss the MinMI criterion.

6.1 System Identification Under the MinMI Criterion

The basic idea behind the MinMI criterion is that the model parameters should be determined such that the identification error contains as little as possible information about the input signal. The scheme of this identification method is shown in Figure 6.1. The objective function is the mutual information between the error and the input, and the optimal parameter is solved as

$$W^*_{\text{MinMI}} = \arg\min_{W \in \Omega_W} I(e_k; X_k) \tag{6.1}$$

---

[1] The minimum mutual information rate criterion was also proposed in [124], which minimizes the mutual information rate between the error signal and a certain white Gaussian process (see Appendix I).

System Parameter Identification. DOI: http://dx.doi.org/10.1016/B978-0-12-404574-3.00006-3
© 2013 Tsinghua University Press Ltd. Published by Elsevier Inc. All rights reserved.

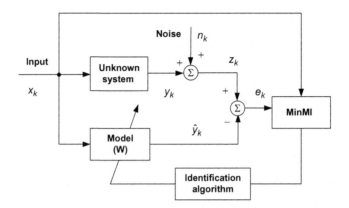

Figure 6.1 System identification under the MinMI criterion.

where e_k is the identification error at k time (the difference between the measurement z_k and the model output $\hat{y}_k$), X_k is a vector consisting of all the inputs that have influence on the model output $\hat{y}_k$ (possibly an infinite dimensional vector), $\Omega_W \subset \mathbb{R}^m$ is the set of all possible m-dimensional parameter vectors.

For a general causal system, X_k will be

$$X_k = [x_k, x_{k-1}, x_{k-2}, \ldots]^T \tag{6.2}$$

If the model output depends on finite input (e.g., the finite impulse response (FIR) filter), then

$$X_k = [x_k, x_{k-1}, \ldots, x_{k-m+1}]^T \tag{6.3}$$

Assume the initial state of the model is known, the output $\hat{y}_k$ will be a function of X_k, i.e., $\hat{y}_k = f(X_k)$. In this case, the MinMI criterion is equivalent to the MEE criterion. In fact, we can derive

$$
\begin{aligned}
W^*_{\text{MinMI}} &= \arg\min_{W \in \Omega_W} I(e_k; X_k) \\
&= \arg\min_{W \in \Omega_W} \{ H(e_k) - H(e_k | X_k) \} \\
&= \arg\min_{W \in \Omega_W} \{ H(e_k) - H(z_k - \hat{y}_k | X_k) \} \\
&= \arg\min_{W \in \Omega_W} \{ H(e_k) - H(z_k - f(X_k) | X_k) \} \\
&= \arg\min_{W \in \Omega_W} \{ H(e_k) - H(z_k | X_k) \} \\
&\overset{(a)}{=} \arg\min_{W \in \Omega_W} H(e_k)
\end{aligned}
\tag{6.4}
$$

where (a) is because that the conditional entropy $H(z_k|X_k)$ is not related to the parameter vector W. In Chapter 3, we have proved a similar property when discussing the MEE Bayesian estimation. That is, minimizing the estimation error entropy is equivalent to minimizing the mutual information between the error and the observation.

Although both are equivalent, the MinMI criterion and the MEE criterion are much different in meaning. The former aims to decrease the statistical dependence while the latter tries to reduce the uncertainty (scatter or dispersion).

6.1.1 Properties of MinMI Criterion

Let the model be an FIR filter. We discuss in the following the optimal solution of the MinMI criterion and investigate the connection to the mean square error (MSE) criterion [234].

Theorem 6.1 For system identification scheme of Figure 6.1, if the model is an FIR filter ($\hat{y}_k = W^T X_k$), z_k and X_k are zero-mean and jointly Gaussian, and the input covariance matrix $R_X \triangleq E[X_k X_k^T]$ satisfies $\det R_X \neq 0$, then we have $W_{\text{MinMI}}^* = W_{\text{MSE}}^* = R_X^{-1} P$, and $I(e_k; X_k)_{W=W_{\text{MinMI}}^*} = 0$, where $P \triangleq E[X_k z_k]$, W_{MSE}^* denotes the optimal weight vector under MSE criterion.

Proof: According to the mean square estimation theory [235], $W_{\text{MSE}}^* = R_X^{-1} P$, thus we only need to prove $W_{\text{MinMI}}^* = R_X^{-1} P$. As $\hat{y}_k = W^T X_k$, we have

$$
\begin{aligned}
E[e_k^2] &= E[(z_k - W^T X_k)(z_k - W^T X_k)^T] \\
&= W^T E[X_k X_k^T]W - 2E[z_k X_k^T]W + E[z_k^2] \\
&= W^T R_X W - 2P^T W + \sigma_z^2
\end{aligned} \tag{6.5}
$$

where $\sigma_z^2 = E[z_k^2]$. And then, we can derive the following gradient

$$
\begin{aligned}
\frac{\partial}{\partial W} I(e_k; X_k) &= \frac{\partial}{\partial W} \{H(e_k) - H(e_k|X_k)\} \\[2mm]
&= \frac{\partial}{\partial W} \{H(e_k) - H(z_k - W^T X_k|X_k)\} \\[2mm]
&= \frac{\partial}{\partial W} \{H(e_k) - H(z_k|X_k)\} \\[2mm]
&\overset{(a)}{=} \frac{\partial}{\partial W} \{H(e_k)\} \overset{(b)}{=} \frac{\partial}{\partial W} \left\{ \frac{1}{2} + \frac{1}{2}\log(2\pi E[e_k^2]) \right\} \\[2mm]
&= \frac{1}{2}\frac{\partial}{\partial W} \log\{W^T R_X W - 2P^T W + \sigma_z^2\} \\[2mm]
&= \frac{R_X W - P}{W^T R_X W - 2P^T W + \sigma_z^2}
\end{aligned} \tag{6.6}
$$

where (*a*) follows from the fact that the conditional entropy $H(z_k|X_k)$ does not depend on the weight vector W and (*b*) is because that e_k is zero-mean Gaussian. Let this gradient be zero, we obtain $W^*_{\text{MinMI}} = R_X^{-1}P$. Next we prove $I(e_k; X_k)_{W=W^*_{\text{MinMI}}} = 0$. By (2.28), we have

$$
\begin{aligned}
I(e_k; X_k)_{W=W^*_{\text{MinMI}}} &= \frac{1}{2}\log\left\{\frac{E[e_k^2]\det E(X_k X_k^T)}{\det\begin{pmatrix} E[e_k^2] & E(e_k X_k^T) \\ E(e_k X_k) & E(X_k X_k^T) \end{pmatrix}}\right\} \\
&\overset{(c)}{=} \frac{1}{2}\log\left\{\frac{E[e_k^2]\det E(X_k X_k^T)}{\det\begin{pmatrix} E[e_k^2] & 0 \\ 0 & E(X_k X_k^T) \end{pmatrix}}\right\} = 0
\end{aligned}
\tag{6.7}
$$

where (*c*) follows from $E(e_k X_k^T) = E[(z_k - W^{*T}_{\text{MinMI}}X_k)X_k^T] = 0$.

Theorem 6.1 indicates that with Gaussian assumption, the optimal FIR filter under MinMI criterion will be equivalent to that under the MSE criterion (i.e., the Wiener solution), and the MinMI between the error and the input will be zero.

Theorem 6.2 If the unknown system and the model are both FIR filters with the same order, and the noise signal $\{n_k\}$ is independent of the input sequence $\{x_k\}$ (both can be of arbitrary distribution), then we have $W^*_{\text{MinMI}} = W_0$, where $W_0 \in \mathbb{R}^m$ denotes the weight vector of unknown system.

Proof: Without Gaussian assumption, Theorem 6.1 cannot be applied here. Let $\tilde{W} = W_0 - W$ be the weight error vector between the unknown system and the model. We have $e_k = X_k^T\tilde{W} + n_k$ and

$$
\begin{aligned}
I(e_k; X_k) &= I(X_k^T\tilde{W} + n_k; X_k) \\
&= H(X_k^T\tilde{W} + n_k) - H(X_k^T\tilde{W} + n_k|X_k) \\
&= H(X_k^T\tilde{W} + n_k) - H(n_k|X_k) \\
&\overset{(a)}{\geq} H(n_k) - H(n_k|X_k) \\
&= I(n_k; X_k) = 0
\end{aligned}
\tag{6.8}
$$

where (*a*) is due to the fact that the entropy of the sum of two independent random variables is not less than the entropy of each individual variable. The equality in (*a*) holds if and only if $\tilde{W} = \mathbf{0}$, i.e., $W^*_{\text{MinMI}} = W_0$.

Theorem 6.2 suggests that the MinMI criterion might be robust with respect to the independent additive noise despite its distribution.

Theorem 6.3 Under the conditions of Theorem 6.2, and assuming that the input $\{x_k\}$ and the noise $\{n_k\}$ are both unit-power white Gaussian processes, then

$$I(e_k; X_k) = -\frac{1}{2}\log(\mathrm{MMSE}(\|\tilde{W}\|))\qquad(6.9)$$

where $\quad\mathrm{MMSE}(\|\tilde{W}\|)\triangleq E[\tilde{W}_0^T(X_k-\hat{X}_k)]^2, \quad \tilde{W}_0 = \tilde{W}/\|\tilde{W}\|, \quad \|\tilde{W}\|\triangleq\sqrt{\tilde{W}^T\tilde{W}},$
$\hat{X}_k = E[X_k|e_k].$

Proof: Obviously, we have $\tilde{W} = \|\tilde{W}\|\tilde{W}_0$, and

$$e_k = \|\tilde{W}\|\tilde{W}_0^T X_k + n_k\qquad(6.10)$$

By the mean square estimation theory [235],

$$
\begin{aligned}
&E[(X_k - \hat{X}_k)(X_k-\hat{X}_k)^T]\\
&= I - \|\tilde{W}\|^2\tilde{W}_0\left[1+\|\tilde{W}\|^2\tilde{W}_0^T\tilde{W}_0\right]^{-1}\tilde{W}_0^T\\
&\overset{(a)}{=} I - \frac{\|\tilde{W}\|^2}{1+\|\tilde{W}\|^2}\tilde{W}_0\tilde{W}_0^T
\end{aligned}
\qquad(6.11)
$$

where (a) follows from $\tilde{W}_0^T\tilde{W}_0 = 1$, I is an $m \times m$ identity matrix. Therefore

$$
\begin{aligned}
\mathrm{MMSE}(\|\tilde{W}\|) &= E\left[\|\tilde{W}_0^T(X_k-\hat{X}_k)\|^2\right]\\
&= \tilde{W}_0^T E[(X_k - \hat{X}_k)(X_k-\hat{X}_k)^T]\tilde{W}_0\\
&= \tilde{W}_0^T\left\{I - \frac{\|\tilde{W}\|^2}{1+\|\tilde{W}\|^2}\tilde{W}_0\tilde{W}_0^T\right\}\tilde{W}_0\\
&= \frac{1}{1+\|\tilde{W}\|^2}
\end{aligned}
\qquad(6.12)
$$

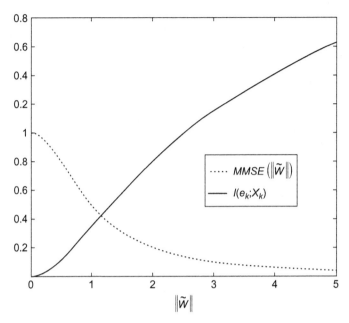

Figure 6.2 Mutual information $I(e_k; X_k)$ and the minimum MSE MMSE$(\|\tilde{W}\|)$ versus weight error norm.
Source: Adopted from [234].

On the other hand, by (2.28), the mutual information $I(e_k; X_k)$ can be calculated as

$$
\begin{aligned}
I(e_k; X_k) &= \frac{1}{2}\log\left\{(1 + \|\tilde{W}\|^2)\det\begin{bmatrix} I & \tilde{W} \\ \tilde{W}^T & \|\tilde{W}\|^2 + 1 \end{bmatrix}^{-1}\right\} \\
&= \frac{1}{2}\log\left(1 + \|\tilde{W}\|^2\right)
\end{aligned}
\tag{6.13}
$$

Combining (6.12) and (6.13) yields the result.

The term MMSE$(\|\tilde{W}\|)$ in Theorem 6.3 is actually the minimum MSE when estimating $\tilde{W}_0^T X_k$ based on e_k. Figure 6.2 shows the mutual information $I(e_k; X_k)$ and the minimum MSE MMSE$(\|\tilde{W}\|)$ versus different weight error norm $\|\tilde{W}\|$. It can be seen that as $\|\tilde{W}\| \to 0$ (or $W \to W_0$), we have $I(e_k; X_k) \to 0$ and MMSE$(\|\tilde{W}\|) \to \max_{\tilde{W}}$ MMSE$(\|\tilde{W}\|)$. This implies that when the model weight vector W approaches the system weight vector W_0, the error e_k contains less and less information about the input vector X_k (or the information contained in the input signal has been sufficiently utilized), and it becomes more and more difficult to estimate the input based on the error signal (i.e., the minimum MSE MMSE$(\|\tilde{W}\|)$ attains gradually its maximum value).

Figure 6.3 General configuration of the ICA.

6.1.2 Relationship with Independent Component Analysis

The parameter identification under MinMI criterion is actually a special case of independent component analysis (ICA) [133]. A brief scheme of the ICA problem is shown in Figure 6.3, where $\vec{s}_k = [s_1(k), s_2(k), \ldots, s_N(k)]^T$ is the N-dimensional source vector, $\vec{x}_k = [x_1(k), x_2(k), \ldots, x_M(k)]^T$ is the M-dimensional observation vector that is related to the source vector through $\vec{x}_k = A\vec{s}_k$, where A is the $M \times N$ mixing matrix [236]. Assume that each component of the source signal $\vec{s}_k$ is mutually independent. There is no other prior knowledge about $\vec{s}_k$ and the mixing matrix A. The aim of the ICA is to search a $N \times M$ matrix B (i.e., the demixing matrix) such that $\vec{y}_k = B\vec{x}_k$ approaches as closely as possible $\vec{s}_k$ up to scaling and permutation ambiguities.

The ICA can be formulated as an optimization problem. To make each component of $\vec{y}_k$ as mutually independent as possible, one can solve the matrix B under a certain objective function that measures the degree of dependence (or independence). Since the mutual information measures the statistical dependence between random variables, we may use the mutual information between components of $\vec{y}_k$ as the optimization criterion,[2] i.e.,

$$B^* = \arg\min_B I(\vec{y}_k) = \arg\min_B \left\{ \sum_{i=1}^{N} H(y_i) - H(\vec{y}_k) \right\} \qquad (6.14)$$

To some extent, the system parameter identification can be regarded as an ICA problem. Consider the FIR system identification:

$$\begin{cases} z_k = W_0^T X_k + n_k \\ \hat{y}_k = W^T X_k \end{cases} \qquad (6.15)$$

where $X_k = [x_k, x_{k-1}, \ldots, x_{k-m+1}]^T$, W_0 and W are m-dimensional weight vectors of the unknown system and the model. If regarding the vectors $[X_k^T, n_k]^T$ and $[X_k^T, z_k]^T$ as, respectively, the source signal and the observation in ICA, we have

$$\begin{bmatrix} X_k \\ z_k \end{bmatrix} = \begin{bmatrix} I & 0 \\ W_0^T & 1 \end{bmatrix} \begin{bmatrix} X_k \\ n_k \end{bmatrix} \qquad (6.16)$$

---

[2] The mutual information minimization is a basic optimality criterion in ICA. Other ICA criteria, such as the negentropy maximization, Infomax, likelihood maximization, and the higher order statistics, in general conform with the mutual information minimization.

where $\begin{bmatrix} I & 0 \\ W_0^T & 1 \end{bmatrix}$ is the mixing matrix and I is the $m \times m$ identity matrix. The goal of the parameter identification is to make the model weight vector W approximate the unknown weight vector W_0, and hence make the identification error e_k ($e_k = z_k - \hat{y}_k$) approach the additive noise n_k, or in other words, make the vector $[X_k^T, e_k]^T$ approach the ICA source vector $[X_k^T, n_k]^T$. Therefore, the vector $[X_k^T, e_k]^T$ can be regarded as the demixing output vector, where the demixing matrix is

$$B = \begin{bmatrix} I & 0 \\ -W^T & 1 \end{bmatrix} \tag{6.17}$$

Due to the scaling ambiguity of the demixing output, it is reasonable to introduce a more general demixing matrix:

$$B = \begin{bmatrix} I & 0 \\ W'^T & a \end{bmatrix} \tag{6.18}$$

where $a \neq 0$, $W' = -aW$. In this case, the demixed output e_k will be related to the identification error via a proportional factor a.

According to (6.14), the optimal demixing matrix will be

$$\begin{aligned} B^* &= \arg \min_B I(e_k; X_k) \\ &= \arg \min_B \{ H(e_k) + H(X_k) - H(e_k, X_k) \} \end{aligned} \tag{6.19}$$

After obtaining the optimal matrix $B^* = \begin{bmatrix} I & 0 \\ W'^{*T} & a^* \end{bmatrix}$, one may get the optimal weight vector [133]

$$W^* = -\frac{W'^*}{a^*} \tag{6.20}$$

Clearly, the above ICA formulation is actually the MinMI criterion-based parameter identification.

6.1.3 ICA-Based Stochastic Gradient Identification Algorithm

The MinMI criterion is in essence equivalent to the MEE criterion. Thus, one can utilize the various information gradient algorithms in Chapter 4 to implement the MinMI criterion-based identification. In the following, we introduce an ICA-based stochastic gradient identification algorithm [133].

According to the previous discussion, the MinMI criterion-based identification can be regarded as an ICA problem, i.e.,

$$B^* = \arg \min_B \{ H(e_k) + H(X_k) - H(e_k, X_k) \} \tag{6.21}$$

where the demixing matrix $B = \begin{bmatrix} I & 0 \\ W'^T & a \end{bmatrix}$.

Since

$$\begin{bmatrix} X_k \\ e_k \end{bmatrix} = B \begin{bmatrix} X_k \\ z_k \end{bmatrix} = \begin{bmatrix} I & 0 \\ W'^T & a \end{bmatrix} \begin{bmatrix} X_k \\ z_k \end{bmatrix} \tag{6.22}$$

by (2.8), we have

$$H(e_k, X_k) = H(z_k, X_k) + \log|a| \tag{6.23}$$

And hence

$$\begin{aligned} B^* &= \arg\min_{B}\{H(e_k) + H(X_k) - [H(z_k, X_k) + \log|a|]\} \\ &= \arg\min_{B}\{[H(e_k) - \log|a|] + [H(X_k) - H(z_k, X_k)]\} \\ &\overset{(a)}{=} \arg\min_{B}\{H(e_k) - \log|a|\} \end{aligned} \tag{6.24}$$

where (a) is due to the fact that the term $[H(X_k) - H(z_k, X_k)]$ is not related to the matrix B. Denote the objective function $J = H(e_k) - \log|a|$. The instantaneous value of J is

$$\hat{J} = -\log p_e(e_k) - \log|a| \tag{6.25}$$

in which $p_e(.)$ is the PDF of e_k ($e_k = W'^T X_k + a z_k$).

In order to solve the demixing matrix B, one can resort to the natural gradient (or relative gradient)-based method [133,237]:

$$\begin{aligned} B_{k+1} &= B_k - \eta \frac{\partial \hat{J}}{\partial B_k} B_k^T B_k \\ &= \begin{bmatrix} I & 0 \\ W_k'^T & a_k \end{bmatrix} - \eta \begin{bmatrix} 0 & 0 \\ \dfrac{\partial \hat{J}}{\partial W_k'^T} & \dfrac{\partial \hat{J}}{\partial a_k} \end{bmatrix} \begin{bmatrix} I + W_k' W_k'^T & W_k' a_k \\ W_k'^T a_k & a_k^2 \end{bmatrix} \\ &= \begin{bmatrix} I & 0 \\ W_k'^T & a_k \end{bmatrix} - \eta \begin{bmatrix} 0 & 0 \\ \varphi(e_k) X_k^T & \varphi(e_k) z_k - \dfrac{1}{a_k} \end{bmatrix} \begin{bmatrix} I + W_k' W_k'^T & W_k' a_k \\ W_k'^T a_k & a_k^2 \end{bmatrix} \\ &= \begin{bmatrix} I & 0 \\ W_k'^T & a_k \end{bmatrix} - \eta \begin{bmatrix} 0 & 0 \\ \varphi(e_k) X_k^T + (\varphi(e_k) e_k - 1) W_k'^T & (\varphi(e_k) e_k - 1) a_k \end{bmatrix} \end{aligned} \tag{6.26}$$

where $\varphi(e_k) = -p'_e(e_k)/p_e(e_k)$. As the PDF $p_e(.)$ is usually unknown, a certain nonlinear function (e.g., the tanh function) will be used to approximate the φ function.[3]

If adopting different step-sizes for learning the parameters W' and a, we have

$$\begin{cases} W'_{k+1} = W'_k + \eta_1[(1 - \varphi(e_k)e_k)W'_k - \varphi(e_k)X_k] \\ a_{k+1} = a_k + \eta_2(1 - \varphi(e_k)e_k)a_k \end{cases} \quad (6.27)$$

The above algorithm is referred to as the ICA-based stochastic gradient identification algorithm (or simply the ICA algorithm). The model weight vector learned by this method is

$$W_k = -\frac{W'_k}{a_k} \quad (6.28)$$

If the parameter a is set to constant $a = 1$, the algorithm will reduce to

$$W'_{k+1} = W'_k - \eta_1 \varphi(e_k)X_k \quad (6.29)$$

6.1.4 Numerical Simulation Example

Figure 6.4 illustrates a general configuration of an acoustic echo canceller (AEC) [133]. x_k is the far-end signal going to the loudspeaker, and y_k is the echo signal entering into the microphone that is produced by an undesirable acoustic coupling between the loudspeaker and the microphone. n_k is the near-end signal which is usually independent of the far-end signal and the echo signal. z_k is the signal received by the microphone ($z_k = y_k + n_k$). The aim of the echo cancelation is to remove the echo part in z_k by subtracting the output of an adaptive filter that is driven by the far-end signal. As shown in Figure 6.4, the filter output $\hat{y}_k$ is the synthetic echo signal, and the error signal e_k is the echo-canceled signal (or the estimate of the near-end signal). The key technique in

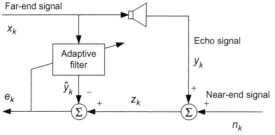

Figure 6.4 General configuration of an AEC.

---

[3] One can also apply the kernel density estimation method to estimate $p_e(.)$, and then use the estimated PDF to compute the φ function.

AEC is to build an accurate model for the echo channel (or accurately identifying the parameters of the synthetic filter).

One may use the previously discussed ICA algorithm to implement the adaptive echo cancelation [133]. Suppose the echo channel is a 100 tap FIR filter, and assume that the input (far-end) signal x_k is uniformly distributed over the interval $[-4, 4]$, and the noise (near-end) signal n_k is Cauchy distributed, i.e., $n_k \sim$ Cauchy(location, scale). The performance of the algorithms is measured by the echo return loss enhancement (ERLE) in dB:

$$\text{ERLE} \triangleq 10 \lg \left(\frac{E[y_k^2]}{E[(y_k - \hat{y}_k)^2]} \right) \tag{6.30}$$

Simulation results are shown in Figures 6.5 and 6.6. In Figure 6.5, the performances of the ICA algorithm, the normalized least mean square (NLMS), and the recursive least squares (RLS) are compared, while in Figure 6.6, the performances of the ICA algorithm and the algorithm (6.29) with $a = 1$ are compared. During the simulation, the φ function in the ICA algorithm is chosen as

$$\varphi(x) = \begin{cases} \tanh(x), & |x| \leq 40 \\ 0 & |x| > 40 \end{cases} \tag{6.31}$$

It can be clearly seen that the ICA-based algorithm shows excellent performance in echo cancelation.

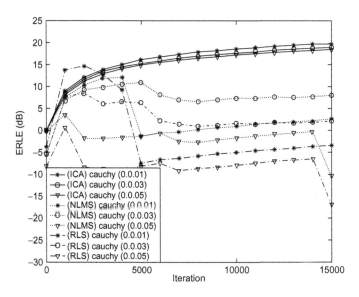

Figure 6.5 Plots of the performance of three algorithms (ICA, NLMS, RLS) in Cauchy noise environment.
Source: Adopted from [133].

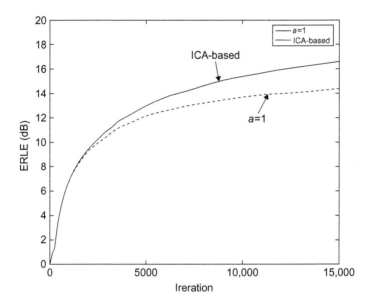

Figure 6.6 Plots of the performance of the ICA algorithm and the algorithm (6.29) with $a = 1$ ($n_k \sim \text{Cauchy}(0, 0.1)$).
Source: Adopted from [133].

6.2 System Identification Under the MaxMI Criterion

Consider the system identification scheme shown in Figure 6.7, in which x_k is the common input to the unknown system and the model, y_k is the intrinsic (noiseless) output of the unknown system, n_k is the additive noise, z_k is the noisy output measurement, and $\hat{y}_k$ stands for the output of the model. Under the MaxMI criterion, the identification procedure is to determine a model M such that the mutual information between the noisy system output z_k and the model output $\hat{y}_k$ is maximized. Thus the optimal model M_{opt} is given by

$$
\begin{aligned}
M_{\text{opt}} &= \arg \max_{M \in \mathbf{M}} I(z_k; \hat{y}_k) \\
&= \arg \max_{M \in \mathbf{M}} \left\{ \int p_{z\hat{y}}(\xi, \tau) \log \frac{p_{z\hat{y}}(\xi, \tau)}{p_z(\xi) p_{\hat{y}}(\tau)} \, d\xi \, d\tau \right\}
\end{aligned}
\tag{6.32}
$$

where $\mathbf{M}$ denotes the model set (collection of all candidate models), $p_z(.)$, $p_{\hat{y}}(.)$, and $p_{z\hat{y}}(.)$ denote, respectively, the PDFs of z_k, $\hat{y}_k$, and $(z_k, \hat{y}_k)$.

The MaxMI criterion provides a fresh insight into system identification. Roughly speaking, the noisy measurement z_k represents the output of an information source

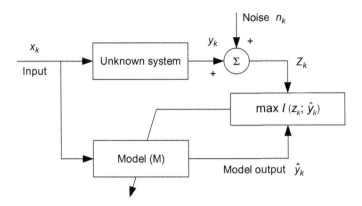

Figure 6.7 Scheme of the system identification under the MaxMI criterion.

and is transmitted over an information channel, i.e., the identifier (including the model set and search algorithm), and the model output $\hat{y}_k$ represents the channel output. Then the identification problem can be regarded as the information transmitting problem, and the goal of identification is to maximize the channel capacity (measured by $I(z_k; \hat{y}_k)$) over all possible identifiers.

6.2.1 Properties of the MaxMI Criterion

In the following, we present some important properties of the MaxMI criterion [135,136].

Property 6.1: Maximizing the mutual information $I(z_k; \hat{y}_k)$ is equivalent to minimizing the conditional error entropy $H(e_k|\hat{y}_k)$, where $e_k = z_k - \hat{y}_k$.

Proof: It is easy to derive

$$
\begin{aligned}
I(z_k; \hat{y}_k) &= H(z_k) - H(z_k|\hat{y}_k) \\
&= H(z_k) - H(z_k - \hat{y}_k|\hat{y}_k) \\
&= H(z_k) - H(e_k|\hat{y}_k)
\end{aligned}
\tag{6.33}
$$

And hence

$$
\begin{aligned}
&\arg\max_{M \in \mathbf{M}} I(z_k; \hat{y}_k) \\
&= \arg\max_{M \in \mathbf{M}} \{H(z_k) - H(e_k|\hat{y}_k)\} \\
&\overset{(a)}{=} \arg\min_{M \in \mathbf{M}} H(e_k|\hat{y}_k)
\end{aligned}
\tag{6.34}
$$

where (a) is due to the fact that the model M has no effect on the entropy $H(z_k)$.

The second property states that under certain conditions, the MaxMI criterion will be equivalent to maximizing the correlation coefficient.

Property 6.2: If z_k and $\hat{y}_k$ are jointly Gaussian, we have $\arg \max_{M \in \mathbf{M}} I(z_k; \hat{y}_k) = \arg \max_{M \in \mathbf{M}} \rho(z_k, \hat{y}_k)$, where $\rho(z_k, \hat{y}_k)$ is the correlation coefficient between z_k and $\hat{y}_k$.

Proof: Since z_k and $\hat{y}_k$ are jointly Gaussian, the mutual information $I(z_k; \hat{y}_k)$ can be calculated as

$$I(z_k; \hat{y}_k) = -\frac{1}{2} \log\{1 - \rho^2(z_k, \hat{y}_k)\} \tag{6.35}$$

The log function is monotonically increasing, thus we have

$$\arg \max_{M \in \mathbf{M}} I(z_k; \hat{y}_k) = \arg \max_{M \in \mathbf{M}} \rho(z_k, \hat{y}_k) \tag{6.36}$$

Property 6.3: Assume the noise n_k is independent of the input signal x_k. Then maximizing the mutual information $I(z_k; \hat{y}_k)$ is equivalent to maximizing a lower bound of the intrinsic (noiseless) mutual information $I(y_k; \hat{y}_k)$.

Proof: Denote $\bar{e}_k$ the intrinsic error, i.e., $\bar{e}_k = y_k - \hat{y}_k$, we have

$$\begin{aligned}
I(z_k; \hat{y}_k) &= H(z_k) - H(z_k|\hat{y}_k) \\
&= H(z_k) - H(y_k + n_k|\hat{y}_k) \\
&= H(z_k) - H(y_k - \hat{y}_k + n_k|\hat{y}_k) \\
&= H(z_k) - H(\bar{e}_k + n_k|\hat{y}_k) \\
&\overset{(b)}{\leq} H(z_k) - H(\bar{e}_k|\hat{y}_k) \\
&= \{H(z_k) - H(y_k)\} + \{H(y_k) - H(\bar{e}_k|\hat{y}_k)\} \\
&= \{H(z_k) - H(y_k)\} + \{H(y_k) - H(y_k - \hat{y}_k|\hat{y}_k)\} \\
&= \{H(z_k) - H(y_k)\} + I(y_k; \hat{y}_k)
\end{aligned} \tag{6.37}$$

where (b) follows from the independence condition and the fact that the entropy of the sum of two independent random variables is not less than the entropy of each individual variable. It follows easily that

$$I(y_k; \hat{y}_k) \geq I(z_k; \hat{y}_k) - \{H(z_k) - H(y_k)\} \tag{6.38}$$

which completes the proof.

In Figure 6.7, the measurement z_k may be further distorted by a certain function. Denote $\tilde{z}_k$ the distorted measurement, we have

$$\tilde{z}_k = \beta[z_k] = \beta[y_k + n_k] \tag{6.39}$$

where $\beta(.)$ is the distortion function. Such distortion widely exists in practical systems. Typical examples include the saturation and the dead zone.

Property 6.4: Suppose the noisy measurement z_k is distorted by a function $\beta(.)$. Then maximizing the distorted mutual information, $I(\beta(z_k); \hat{y}_k)$ is equivalent to maximizing a lower bound of the undistorted mutual information $I(z_k; \hat{y}_k)$.

Proof: This property is a direct consequence of the *data processing inequality* (see Theorem 2.3), which states that for any random variables X and Y, and any measurable function $\beta(.)$,

$$I(\beta(X); Y) \leq I(X; Y) \tag{6.40}$$

In (6.40), if function $\beta(.)$ is invertible, the equality will hold. In this case, we have

$$\arg \max_{M \in \mathbf{M}} I(\beta(z_k); \hat{y}_k) = \arg \max_{M \in \mathbf{M}} I(z_k; \hat{y}_k) \tag{6.41}$$

That is, the invertible distortion does not change the optimal solutions of MaxMI.

Property 6.5: If the measurement z_k is Gaussian, then maximizing the mutual information $I(z_k; \hat{y}_k)$ will be equivalent to minimizing a lower bound of the MSE.

Proof: According to Theorem 2.4, the rate distortion function for a Gaussian source $X \sim \mathcal{N}(\mu, \sigma^2)$ with MSE distortion is

$$R(D) = \frac{1}{2} \log \frac{\sigma^2}{D^2}, \quad D \geq 0 \tag{6.42}$$

where $R(D) \triangleq \inf_Y \{I(X; Y) : E[(X - Y)^2] \leq D^2\}$. Let $X = z_k$, we have

$$\begin{aligned} I(z_k; \hat{y}_k) &\geq \inf_Y \{I(z_k; Y); E[(z_k - Y)^2] \leq E[(z_k - \hat{y}_k)^2]\} \\ &= R(E[(z_k - \hat{y}_k)^2]) \\ &= \frac{1}{2} \log \frac{\sigma_z^2}{E[(z_k - \hat{y}_k)^2]} \end{aligned} \tag{6.43}$$

where σ_z^2 is the variance of z_k. It follows easily that

$$
\begin{aligned}
E[e_k^2] &= E[(z_k - \hat{y}_k)^2] \\
&\geq \sigma_z^2 \exp(-2I(z_k; \hat{y}_k))
\end{aligned}
\tag{6.44}
$$

This completes the proof.

Consider now a special case where the model is represented by an FIR filter in which the output $\hat{y}_k$ is given by

$$
\hat{y}_k = X_k^T W
\tag{6.45}
$$

where $X_k = [x_k, x_{k-1}, \ldots, x_{k-m+1}]^T$ is the input (regressor) vector and $W = [w_0, w_1, \ldots, w_{m-1}]^T$ is the weight vector. Then we have the following results.

Property 6.6: For the case of the FIR model and under the assumption that z_k and X_k are jointly Gaussian, the optimal weight vector under the MaxMI criterion will be

$$
W_{\mathrm{opt}} = \arg \max_{W \in \mathbb{R}^m} I(z_k; \hat{y}_k) = \gamma R_X^{-1} P
\tag{6.46}
$$

where $R_X = E[(X_k - E[X_k])(X_k - E[X_k])^T]$, $P = E[(X_k - E[X_k])(z_k - E[z_k])]$, $\gamma \in \mathbb{R}$ ($\gamma \neq 0$). and in particular, if $\gamma = 1$, the MSE $E[e_k^2]$ will attain the lower bound as in (6.44), i.e.,

$$
E[e_k^2] = \sigma_z^2 \exp(-2I(z_k; \hat{y}_k))
\tag{6.47}
$$

Proof: Since z_k and X_k are jointly Gaussian, then z_k and $\hat{y}_k$ are also jointly Gaussian. By Property 6.2, we have

$$
\begin{aligned}
W_{\mathrm{opt}} &= \arg \max_{W \in \mathbb{R}^m} \rho(z_k, \hat{y}_k) \\
&= \arg \max_{W \in \mathbb{R}^m} \frac{E[(\hat{y}_k - E[\hat{y}_k])(z_k - E[z_k])]}{\sigma_z \sqrt{E[(\hat{y}_k - E[\hat{y}_k])^2]}} \\
&= \arg \max_{W \in \mathbb{R}^m} \frac{W^T P}{\sigma_z \sqrt{(W^T R_X W)}} \\
&\overset{(c)}{=} \arg \max_{W \in \mathbb{R}^m} \frac{W^T P}{\sqrt{(W^T R_X W)}}
\end{aligned}
\tag{6.48}
$$

where (c) is because that σ_z is not related to W. And then,

$$\frac{\partial}{\partial W} \frac{W^T P}{\sqrt{W^T R_X W}} = \left(\frac{W^T P}{(W^T R_X W)^{3/2}}\right) \left\{\left(\frac{W^T R_X W}{W^T P}\right)P - R_X W\right\} \tag{6.49}$$

Let the above gradient be zero, and denote $\gamma = (W^T R_X W)/(W^T P)$, we obtain the optimal weight vector

$$W_{\text{opt}} = \gamma R_X^{-1} P \tag{6.50}$$

It can be easily verified that for any $\gamma \in \mathbb{R}$, and $\gamma \neq 0$, the optimal weight vector (6.50) makes the gradient (6.49) zero. When $\gamma = 1$, the optimal weight becomes the Wiener solution $W_{\text{opt}} = R_X^{-1} P$. In this case, the MSE is

$$\begin{aligned}
E[e_k^2] &= W^T R_X W - 2P^T W + \sigma_z^2 \\
&= \sigma_z^2 - P^T R_X^{-1} P
\end{aligned} \tag{6.51}$$

Further the mutual information $I(z_k; \hat{y}_k)$ is

$$\begin{aligned}
I(z_k; \hat{y}_k) &= -\frac{1}{2}\log\{1 - \rho^2(z_k, \hat{y}_k)\} \\
&= -\frac{1}{2}\log\left\{1 - \frac{\left(E[(\hat{y}_k - E[\hat{y}_k])(z_k - E[z_k])]\right)^2}{E[(\hat{y}_k - E[\hat{y}_k])^2]\sigma_z^2}\right\} \\
&= -\frac{1}{2}\log\left\{1 - \frac{(W^T P)^2}{W^T R_X W \sigma_z^2}\right\} \\
&= -\frac{1}{2}\log\left\{1 - \frac{P^T R_X^{-1} P}{\sigma_z^2}\right\}
\end{aligned} \tag{6.52}$$

Combining (6.51) and (6.52), we obtain $E[e_k^2] = \sigma_z^2 \exp(-2I(z_k; \hat{y}_k))$.

Property 6.6 indicates that with a FIR filter structure and under Gaussian assumption, the MaxMI criterion yields a scaled Wiener solution which is not unique. Thus it does not satisfy the identifiability condition.[4] The main reason for this is that any invertible transformation does not change the mutual information. In this property, γ is restricted to nonzero. If $\gamma = 0$, we have $W_{\text{opt}} = \mathbf{0}$, and the mutual information $I(z_k; \hat{y}_k) = I(z_k; 0)$ will be undefined (ill-posed).

---

[4] It is worth noting that the identifiability problem under the MaxMI criterion has been studied in [134], wherein the "identifiability" does not means the uniqueness of the solution, but just means that the mutual information between the system output and the model output is nonzero.

A *priori* information usually has great value in system identification. For example, if the structures of the system or the parameters are partially known, we may use this information to impose some constraints on the structures or parameters of the filter. For the case in which the desired responses are distorted, the *a priori* information can help to improve the accuracy of the solution. In particular, certain parameter constraints may yield a unique optimal solution under the MaxMI criterion. Consider the optimal solution (6.50) under the following parameter constraint:

$$C^T W = \alpha \tag{6.53}$$

where $C = [c_1, c_2, \ldots, c_m]^T \in \mathbb{R}^m$, $\alpha \in \mathbb{R}$. Let $W = \gamma R_X^{-1} P$, we have

$$C^T W = \gamma C^T R_X^{-1} P = \alpha \tag{6.54}$$

If $C^T R_X^{-1} P \neq 0$, then γ can be uniquely determined as $\gamma = (C^T R_X^{-1} P)^{-1} \alpha$.

6.2.2 Stochastic Mutual Information Gradient Identification Algorithm

The stochastic gradient identification algorithm under the MaxMI criterion can be expressed as

$$W_{k+1} = W_k + \eta \hat{\nabla}_W I(z_k; \hat{y}_k) \tag{6.55}$$

where $\hat{\nabla}_W I(z_k; \hat{y}_k)$ denotes the instantaneous estimate of the gradient of mutual information $I(z_k; \hat{y}_k)$ evaluated at the current value of the weight vector and η is the step-size. The key problem of the update equation (6.55) is how to calculate the instantaneous gradient $\hat{\nabla}_W I(z_k; \hat{y}_k)$.

Let us start with the calculation of the gradient (not the instantaneous gradient) of $I(z_k; \hat{y}_k)$:

$$\nabla_W I(z_k; \hat{y}_k) = \frac{\partial}{\partial W} I(z_k; \hat{y}_k)$$

$$= \frac{\partial}{\partial W} E \left\{ \log \left(\frac{p_{z\hat{y}}(z_k, \hat{y}_k)}{p_z(z_k) p_{\hat{y}}(\hat{y}_k)} \right) \right\} \tag{6.56}$$

$$= \frac{\partial}{\partial W} E\{\log p_{z\hat{y}}(z_k, \hat{y}_k) - \log p_{\hat{y}}(\hat{y}_k)\}$$

where $p_z(.)$, $p_{\hat{y}}(.)$, and $p_{z\hat{y}}(.)$ denote the related PDFs at the instant k. Then the instantaneous value of $\nabla_W I(z_k; \hat{y}_k)$ can be obtained by dropping the expectation operator and plugging in the estimates of the PDFs, i.e.,

$$
\hat{\nabla}_W I(z_k; \hat{y}_k) = \frac{\partial}{\partial W} \{\log \hat{p}_{z\hat{y}}(z_k, \hat{y}_k) - \log \hat{p}_{\hat{y}}(\hat{y}_k)\}
$$

$$
= \frac{\left(\frac{\partial}{\partial W}\right)\hat{p}_{z\hat{y}}(z_k, \hat{y}_k)}{\hat{p}_{z\hat{y}}(z_k, \hat{y}_k)} - \frac{\left(\frac{\partial}{\partial W}\right)\hat{p}_{\hat{y}}(\hat{y}_k)}{\hat{p}_{\hat{y}}(\hat{y}_k)}
$$

(6.57)

where $\hat{p}_{z\hat{y}}(z_k, \hat{y}_k)$ and $\hat{p}_{\hat{y}}(\hat{y}_k)$ are, respectively, the estimates of $p_{z\hat{y}}(z_k, \hat{y}_k)$ and $p_{\hat{y}}(\hat{y}_k)$. To estimate the density functions, one usually adopts the kernel density estimation (KDE) method and uses the following Gaussian functions as the kernels [135]

$$
\begin{cases}
K_{h_1}(x) = \dfrac{1}{\sqrt{2\pi}h_1} \exp\left(-\dfrac{x^2}{2h_1^2}\right) \\[4mm]
K_{h_2}(x, y) = \dfrac{1}{2\pi h_2^2} \exp\left(-\dfrac{x^2+y^2}{2h_2^2}\right)
\end{cases}
$$

(6.58)

where h_1 and h_2 denote the kernel widths.

Based on the above Gaussian kernels, the estimates of the PDFs and their gradients can be calculated as follows:

$$
\begin{cases}
\hat{p}_{\hat{y}}(\hat{y}_k) = \dfrac{1}{\sqrt{2\pi}Lh_1} \sum_{j=0}^{L-1} \exp\left(-\dfrac{(\hat{y}_k-\hat{y}_{k-j})^2}{2h_1^2}\right) \\[4mm]
\hat{p}_{z\hat{y}}(z_k, \hat{y}_k) = \dfrac{1}{2\pi Lh_2^2} \sum_{j=0}^{L-1} \exp\left(-\dfrac{(z_k-z_{k-j})^2 + (\hat{y}_k-\hat{y}_{k-j})^2}{2h_2^2}\right)
\end{cases}
$$

(6.59)

$$
\begin{cases}
\dfrac{\partial}{\partial W}\hat{p}_{\hat{y}}(\hat{y}_k) = \dfrac{-1}{\sqrt{2\pi}Lh_1^3} \sum_{j=0}^{L-1} \left\{\exp\left(-\dfrac{[\hat{y}_k-\hat{y}_{k-j}]^2}{2h_1^2}\right)\pi_{kj}\right\} \\[4mm]
\dfrac{\partial}{\partial W}\hat{p}_{z\hat{y}}(z_k, \hat{y}_k) = \dfrac{-1}{2\pi Lh_2^4} \sum_{j=0}^{L-1} \left\{\exp\left(-\dfrac{[z_k-z_{k-j}]^2 + [\hat{y}_k-\hat{y}_{k-j}]^2}{2h_2^2}\right)\pi_{kj}\right\}
\end{cases}
$$

(6.60)

where L is the sliding data length and $\pi_{kj} = (\hat{y}_k - \hat{y}_{k-j})(\partial\hat{y}_k/\partial W - \partial\hat{y}_{k-j}/\partial W)$. For FIR filter, we have

$$\pi_{kj} = (\hat{y}_k - \hat{y}_{k-j})(X_k - X_{k-j}) \tag{6.61}$$

Combining (6.55), (6.57), (6.59), and (6.60), we obtain a stochastic gradient identification algorithm under the MaxMI criterion, which is referred to as the stochastic mutual information gradient (SMIG) algorithm [135].

The performances of the SMIG algorithm compared with the least mean square (LMS) algorithm are demonstrated in the following by Monte Carlo simulations. Consider the FIR system identification [135]:

$$\begin{cases} G^*(z) = 0.8 + 0.2z^{-1} + 0.4z^{-2} + 0.6z^{-3} + 0.4z^{-4} + 0.2z^{-5} \\ G(z) = w_0 + w_1 z^{-1} + w_2 z^{-2} + w_3 z^{-3} + w_4 z^{-4} + w_5 z^{-5} \end{cases} \tag{6.62}$$

where $G^*(z)$ and $G(z)$ are, respectively, the transfer functions of the unknown system and the model. Suppose the input signal $\{x_k\}$ and the additive noise $\{n_k\}$ are both unit-power white Gaussian processes. To uniquely determine an optimal solution under the MaxMI criterion, it is assumed that the first component of the unknown weight vector is *a priori* known (which is assumed to be 0.8). Thus the goal is to search the optimal solution of the other five weights. The initial weights (except $w_0 = 0.8$) for the adaptive FIR filter are zero-mean Gaussian distributed with variance 0.01. Further, the following distortion functions are considered [135]:

1. Undistorted: $\beta(x) = x, \quad x \in \mathbb{R}$

2. Saturation: $\beta(x) = \begin{cases} x & x \in [-1, 1] \\ 2 - e^{(1-x)} & x \in (1, +\infty) \\ -2 + e^{(1+x)} & x \in (-\infty, -1) \end{cases}$

3. Dead zone: $\beta(x) = \begin{cases} 0 & x \in [-0.4, 0.4] \\ x - 0.4 & x \in (0.4, +\infty) \\ x + 0.4 & x \in (-\infty, -0.4) \end{cases}$

4. Data loss[5] : $\beta(x): \begin{cases} \Pr\{\beta(x) = 0\} = 0.3 \\ \Pr\{\beta(x) = x\} = 0.7 \end{cases} \quad x \in \mathbb{R}$

Figure 6.8 plots the distortion functions of the saturation and dead zone. Figure 6.9 shows the desired response signal with data loss (the probability of data loss is 0.3). In the simulation below, the Gaussian kernels are used and the kernel sizes are kept fixed at $h_1 = h_2 = 0.4$.

Figure 6.10 illustrates the average convergence curves over 50 Monte Carlo simulation runs. One can see that, without measurement distortions, the conventional LMS algorithm has a better performance. However, in the case of measurement

---

[5] Data loss means that there exists accidental loss of the measurement data due to certain failures in the sensors or communication channels.

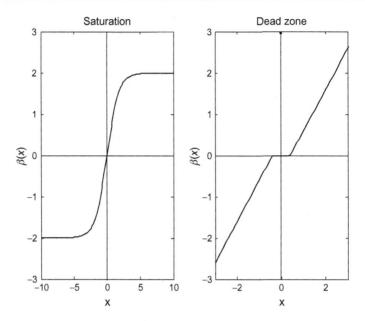

Figure 6.8 Distortion functions of saturation and dead zone.
Source: Adopted from [135].

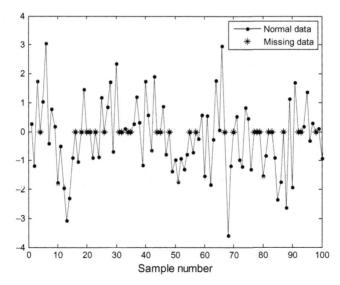

Figure 6.9 Desired response signal with data loss.
Source: Adopted from [135].

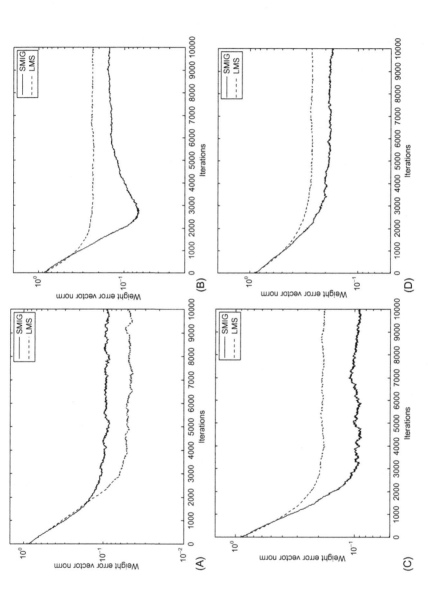

Figure 6.10 Average convergence curves of SMIG and LMS algorithms: (A) undistorted, (B) saturation, (C) dead zone, and (D) data loss. *Source*: Adopted from [135].

Figure 6.11 Wiener system.

distortions, it is evident the deterioration of the LMS algorithm whereas the SMIG algorithm is little affected and achieves a much better performance. Simulation results confirm that the MaxMI criterion is more robust to the measurement distortion than traditional MSE criterion.

6.2.3 Double-Criterion Identification Method

The system identification scheme of Figure 6.7 does not, in general, satisfy the condition of parameter identifiability (i.e., the uniqueness of the optimal solution). In order to uniquely determine an optimal solution, some *a priori* information about the parameters is required. However, such *a priori* information is not available for many practical applications. To address this problem, we introduce in the following the double-criterion identification method [136].

Consider the Wiener system shown in Figure 6.11, where the system has the cascade structure and consists of a discrete-time linear filter $H(z)$ followed by a zero-memory nonlinearity $f(.)$. Wiener systems are typical nonlinear systems and are widely used for nonlinear modeling [238]. The double-criterion method mainly aims at the Wiener system identification, but it also applies to many other systems. In fact, any system can be regarded as a cascade system consisting of itself followed by $f(x) = x$.

First, we define the equivalence between two Wiener systems.

Definition 6.1 Two Wiener systems $\{H_1(z), f_1(.)\}$ and $\{H_2(z), f_2(.)\}$ are said to be equivalent if and only if $\exists \gamma \neq 0$, such that

$$\begin{cases} H_2(z) = \gamma H_1(z) \\ f_2(x) = f_1(x/\gamma) \end{cases} \tag{6.63}$$

Although there is a scale factor γ between two equivalent Wiener systems, they have exactly the same input−output behavior.

The optimal solution of the system identification scheme of Figure 6.7 is usually nonunique. For Wiener system, the nonuniqueness means the optimal solutions are not all equivalent. According to the data processing inequality, we have

$$I(z_k; \hat{y}_k) = I(z_k; \hat{f}(\hat{u}_k)) \leq I(z_k; \hat{u}_k) \tag{6.64}$$

where $\hat{u}_k$ and $\hat{f}(.)$ denote, respectively, the intermediate output (the output of the linear part) and the zero-memory nonlinearity of the Wiener model. Then under the MaxMI criterion, the optimal Wiener model will be

$$\begin{cases} W_{opt} = \arg\max_{W} I(z_k; \hat{u}_k) \\ f_{opt} \in \{\hat{f} \in \mathcal{F} \,|\, I(z_k; \hat{f}(\hat{u}_k)) = I(z_k; \hat{u}_k)\} \end{cases} \tag{6.65}$$

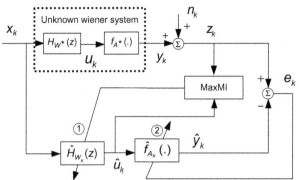

Figure 6.12 Double-criterion identification scheme for Wiener system: (1) linear filter part is identified using MaxMI criterion and (2) nonlinear part is trained using MSE criterion.

where W denotes the parameter vector of the linear subsystem and $\mathscr{F}$ denotes all measurable mappings $\hat{u}_k \to \hat{y}_k$. Evidently, the optimal solutions given by (6.65) contain infinite nonequivalent Wiener models. Actually, we always have $I(z_k; \hat{f}(\hat{u}_k)) = I(z_k; \hat{u}_k)$ provided $\hat{f}$ is an invertible function.

In order to ensure that all the optimal Wiener models are equivalent, the identification scheme of Figure 6.7 has to be modified. One can adopt the double-criterion identification method [136]. As shown in Figure 6.12, the double-criterion method utilizes both MaxMI and MSE criteria to identify the Wiener system. Specifically, the linear filter part is identified by using the MaxMI criterion, and the zero-memory nonlinear part is learned by the MSE criterion. In Figure 6.12, $H_{W^*}(z)$ and $f_{A^*}(.)$ denote, respectively, the linear and nonlinear subsystems of the unknown Wiener system, where W^* and A^* are related parameter vectors. The adaptive Wiener model $\{\hat{H}_{W_k}(z), \hat{f}_{A_k}(.)\}$ usually takes the form of "FIR + polynomial", that is, the linear subsystem $\hat{H}_{W_k}(z)$ is an $(m-1)$-order FIR filter, and the nonlinear subsystem $\hat{f}_{A_k}(.)$ is a $(p-1)$-order polynomial. In this case, the intermediate output $\hat{u}_k$ and the final output $\hat{y}_k$ of the model are

$$
\begin{cases}
\hat{u}_k = W_k^T X_k \\
\hat{y}_k = \hat{f}_{A_k}(\hat{u}_k) = A_k^T \hat{U}_k
\end{cases}
\tag{6.66}
$$

where W_k and X_k are m-dimensional FIR weight vector and input vector, $A_k = [a_0(k), a_1(k), \ldots, a_{p-1}(k)]^T$ is the p-dimensional polynomial coefficient vector, and $\hat{U}_k = [1, \hat{u}_k, \hat{u}_k^2, \ldots, \hat{u}_k^{p-1}]^T$ is the polynomial basis vector.

It should be noted that similar two-gradient identification algorithms for the Wiener system have been proposed in [239,240], wherein the linear and nonlinear subsystems are both identified using the MSE criterion.

The optimal solution for the above double-criterion identification is

$$
\begin{cases}
W_{\text{opt}} = \arg\max_{W \in \mathbb{R}^m} I(z_k; \hat{u}_k) \\
A_{\text{opt}} = \arg\min_{A \in \mathbb{R}^p, W = W_{\text{opt}}} E[(z_k - \hat{y}_k)^2]
\end{cases}
\tag{6.67}
$$

For general case, it is hard to find the closed-form expressions for W_{opt} and A_{opt}. The following theorem only considers the case in which the unknown Wiener system has the same structure as the assumed Wiener model.

Theorem 6.4 For the Wiener system identification scheme shown in Figure 6.12, if

1. The unknown system and the model have the same structure, that is, $H_{W^*}(z)$ and $\hat{H}_W(z)$ are both $(m-1)$-order FIR filters, and $f_{A^*}(.)$ and $\hat{f}_A(.)$ are both $(p-1)$-order polynomials.
2. The nonlinear function $f_{A^*}(.)$ is invertible.
3. The additive noise n_k is independent of the input vector X_k.

Then the optimal solution of (6.67) will be

$$W_{\text{opt}} = \gamma W^*, \quad A_{\text{opt}} = A^* G_\gamma^{-1} \tag{6.68}$$

where $\gamma \in \mathbb{R}$, $\gamma \neq 0$, and G_γ is expressed as

$$G_\gamma = \begin{bmatrix} 1 & 0 & \cdots & 0 \\ 0 & \gamma^1 & \ddots & \vdots \\ \vdots & \ddots & \ddots & 0 \\ 0 & \cdots & 0 & \gamma^{p-1} \end{bmatrix} \tag{6.69}$$

Proof: Since X_k and n_k are mutually independent, we have

$$
\begin{aligned}
I(z_k; \hat{u}_k) &= H(z_k) - H(z_k|\hat{u}_k) \\
&= H(z_k) - H(y_k + n_k|\hat{u}_k) \\
&\overset{(d)}{\leq} H(z_k) - H(n_k|\hat{u}_k) \\
&= H(z_k) - H(n_k)
\end{aligned} \tag{6.70}
$$

where (d) follows from the fact that the entropy of the sum of two independent random variables is not less than the entropy of each individual variable.

In (6.70), the equality holds if and only if conditioned on $\hat{u}_k$, y_k is a deterministic variable, that is, y_k is a function of $\hat{u}_k$. This implies that the mutual information $I(z_k; \hat{u}_k)$ will achieve its maximum value $(H(z_k) - H(n_k))$ if and only if there exists a function $\varphi(.)$ such that $y_k = \varphi(\hat{u}_k)$, i.e.,

$$y_k = f_{A^*}(W^{*T}X_k) = \varphi(\hat{u}_k) = \varphi(W_{\text{opt}}^T X_k), \ \forall X_k \in \mathbb{R}^m \tag{6.71}$$

As the nonlinear function $f_{A^*}(.)$ is assumed to be invertible, we have

$$W^{*T}X_k = f_{A^*}^{-1}(\varphi(W_{opt}^T X_k)) = \psi(W_{opt}^T X_k), \quad \forall X_k \in \mathbb{R}^m \tag{6.72}$$

where $f_{A^*}^{-1}(.)$ denotes the inverse function of $f_{A^*}(.)$ and $\psi \triangleq f_{A^*}^{-1} \circ \varphi$. It follows that

$$\psi(1) = \psi\left(\frac{W_{opt}^T X_k}{W_{opt}^T X_k}\right) = \psi\left(W_{opt}^T \frac{X_k}{(W_{opt}^T X_k)}\right) = W^{*T}\frac{X_k}{(W_{opt}^T X_k)}, \quad \forall X_k \in \mathbb{R}^m \tag{6.73}$$

And hence

$$\psi(1)W_{opt}^T X_k = W^{*T}X_k, \quad \forall X_k \in \mathbb{R}^m \tag{6.74}$$

which implies $\psi(1)W_{opt}^T = W^{*T}$. Let $\gamma = 1/\psi(1) \neq 0$, we obtain the optimum FIR weight $W_{opt} = \gamma W^*$.

Now the optimal polynomial coefficients can be easily determined. By independent assumption, we have

$$E[(z_k - \hat{y}_k)^2] = E[(y_k + n_k - \hat{y}_k)^2] \geq E[n_k^2] \tag{6.75}$$

with equality if and only if $y_k - \hat{y}_k \equiv 0$. This means the MSE cost will attain its minimum value ($E[n_k^2]$) if and only if the intrinsic error ($\bar{e}_k = y_k - \hat{y}_k$) remains zero. Therefore,

$$\begin{aligned}
\bar{e}_k\big|_{W=W_{opt}, A=A_{opt}} &= (y_k - \hat{y}_k)\big|_{W=W_{opt}, A=A_{opt}} \\
&= f_{A^*}(W^{*T}X_k) - \hat{f}_{A_{opt}}(W_{opt}^T X_k) \\
&= f_{A^*}(W^{*T}X_k) - f_{A_{opt}}(\gamma W^{*T}X_k) \\
&= f_{A^*}(W^{*T}X_k) - f_{A_{opt}G_\gamma}(W^{*T}X_k) \\
&\equiv 0
\end{aligned} \tag{6.76}$$

Then we get $A_{opt}G_\gamma = A^* \Rightarrow A_{opt} = A^*G_\gamma^{-1}$, where G_γ is given by (6.69). This completes the proof.

Theorem 6.4 indicates that for identification scheme of Figure 6.12, under certain conditions the optimal solution will match the true system exactly (i.e., with zero intrinsic error). There is a free parameter γ in the solution, however, its specific value has no substantial effect on the cascaded model. The literature [240] gives a similar result about the optimal solution under the single MSE criterion. In [240], the linear FIR subsystem is estimated up to a scaling factor which equals the derivative of the nonlinear function around a bias point.

The double-criterion identification can be implemented in two manners. The first is the sequential identification scheme, in which the MaxMI criterion is first used to learn the linear FIR filter. At the end of the first adaptation phase, the tap weights are frozen, and then the MSE criterion is used to estimate the polynomial coefficients. The second adaptation scheme simultaneously trains both the linear and nonlinear parts of the Wiener model. Obviously, the second scheme is more suitable for online identification. In the following, we focus only on the simultaneous scheme.

In [136], a stochastic gradient-based double-criterion identification algorithm was developed as follows:

$$
\begin{cases}
W_{k+1} = W_k + \eta_1 \hat{\nabla}_W I(z_k; \hat{u}_k) & \text{(a1)} \\
W_{k+1} = \dfrac{W_{k+1}}{\|W_{k+1}\|} & \text{(a2)} \\
A_{k+1} = A_k + \dfrac{\eta_2 e_k \hat{U}_k}{\|\hat{U}_k\|^2} & \text{(a3)}
\end{cases}
\tag{6.77}
$$

where $\hat{\nabla}_W I(z_k; \hat{u}_k)$ denotes the stochastic (instantaneous) gradient of the mutual information $I(z_k; \hat{u}_k)$ with respect to the FIR weight vector W (see (6.57) for the computation), $\|.\|$ denotes the Euclidean norm, η_1 and η_2 are the step-sizes. The update equation (a1) is actually the SMIG algorithm developed in 6.2.2. The second part (a2) of the algorithm (6.77) scales the FIR weight vector to a unit vector. The purpose of scaling the weight vector is to constrain the output energy of the FIR filter, and to avoid "very large values" of the scale factor γ in the optimal solution. As mutual information is scaling invariant,[6] the scaling (a2) does not influence the search of the optimal solution. However, it certainly affects the value of γ in the optimal solution. In fact, if the algorithm converges to the optimal solution, we have $\lim_{k \to \infty} W_k = \gamma W^*$, and

$$
\lim_{k \to \infty} W_{k+1} = \lim_{k \to \infty} \frac{W_{k+1}}{\|W_{k+1}\|}
$$

$$
\Rightarrow \gamma W^* = \frac{\gamma W^*}{\|\gamma W^*\|} = sign(\gamma) \frac{W^*}{\|W^*\|}
\tag{6.78}
$$

$$
\Rightarrow \gamma sign(\gamma) = 1 / \|W^*\|
$$

$$
\Rightarrow \gamma = \pm 1 / \|W^*\|
$$

That is, the scale factor γ equals either $1/\|W^*\|$ or $-1/\|W^*\|$, which is no longer a free parameter.

---

[6] For any random variables X and Y, the mutual information $I(X; Y)$ satisfies $I(X; Y) = I(aX; bY)$, $\forall a \neq 0, b \neq 0$.

The third part (a3) of the algorithm (6.77) is the NLMS algorithm, which minimizes the MSE cost with step-size scaled by the energy of polynomial regression signal $\hat{U}_k$. The NLMS is more suitable for the nonlinear subsystem identification than the standard LMS algorithm, because during the adaptation, the polynomial regression signals are usually nonstationary. The algorithm (6.77) is referred to as the SMIG-NLMS algorithm [136].

Next, Monte Carlo simulation results are presented to demonstrate the performance of the SMIG-NLMS algorithm. For comparison purpose, simulation results of the following two algorithms are also included.

$$
\begin{cases}
W_{k+1} = W_k + \mu_1 e_k^{(1)} X_k \\
A_{k+1} = A_k + \mu_2 e_k^{(2)} \hat{U}_k / \| \hat{U}_k \|^2
\end{cases}
\tag{6.79}
$$

$$
\begin{cases}
W_{k+1} = W_k - \mu_1' \nabla_W \hat{H}(e_k^{(1)}) \\
A_{k+1} = A_k + \mu_2' e_k^{(2)} \hat{U}_k / \| \hat{U}_k \|^2
\end{cases}
\tag{6.80}
$$

where μ_1, μ_2, μ_1', μ_2' are step-sizes, $e_k^{(1)} \triangleq z_k - W_k^T X_k$, $e_k^{(2)} \triangleq z_k - \hat{f}_{A_k}(W_k^T X_k)$, and $\nabla_W \hat{H}(e_k^{(1)})$ denote the stochastic information gradient (SIG) under Shannon entropy criterion, calculated as

$$
\nabla_W \hat{H}(e_k^{(1)}) = \frac{\sum\limits_{i=k-L+1}^{k} K_h'(e_k^{(1)} - e_i^{(1)})(X_k - X_i)}{\sum\limits_{i=k-L+1}^{k} K_h\left(e_k^{(1)} - e_i^{(1)}\right)}
\tag{6.81}
$$

where $K_h(.)$ denotes the kernel function with bandwidth h. The algorithms (6.79) and (6.80) are referred to as the LMS-NLMS and SIG-NLMS algorithms [136], respectively. Note that the LMS-NLMS algorithm is actually the normalized version of the algorithm developed in [240].

Due to the "scaling" property of the linear and nonlinear portions, the expression of Wiener system is not unique. In order to evaluate how close the estimated Wiener model and the true system are, we introduce the following measures [136]:

1. Angle between W_k and W^*:

$$
\theta(W_k, W^*) = \min\{ \angle (W_k, W^*), \angle (-W_k, W^*) \}
\tag{6.82}
$$

where $\angle (W_k, W^*) = arc \cos\left((W_k^T W^*)/(\| W_k \| \| W^* \|)\right)$.
2. Angle between A_k and A^*:

$$
\theta(A_k, A^*) = \min\{ \angle (A_k, A^*), \angle (A_k G_{-1}, A^*) \}
\tag{6.83}
$$

where G_{-1} is

$$G_{-1} = \begin{bmatrix} (-1)^0 & 0 & \cdots & 0 \\ 0 & (-1)^1 & \ddots & \vdots \\ \vdots & \ddots & \ddots & 0 \\ 0 & \cdots & 0 & (-1)^p \end{bmatrix} \qquad (6.84)$$

3. Intrinsic error power (IEP)[7]:

$$\text{IEP} \triangleq E[\bar{e}_k^2] = E[(y_k - \hat{y}_k)^2] \qquad (6.85)$$

Among the three performance measures, the angles $\theta(W_k, W^*)$ and $\theta(A_k, A^*)$ quantify the identification performance of the subsystems (linear FIR and nonlinear polynomial), while the IEP quantifies the overall performance.

Let us consider the case in which the FIR weights and the polynomial coefficients of the unknown Wiener system are [136]

$$\begin{cases} W^* = [0.3, 0.5, -0.6, -0.2, 0.4, 0.3, 0.1] \\ A^* = [0.2, 1.0, 0.5, 0.1] \end{cases} \qquad (6.86)$$

The common input $\{x_k\}$ is a white Gaussian process with unit variance and the disturbance noise $\{n_k\}$ is another white Gaussian process with variance $\sigma_n^2 = 0.01$. The initial FIR weight vector W_0 of the adaptive model is obtained by normalizing a zero-mean Gaussian-distributed random vector ($\|W_0\| = 1$), and the initial polynomial coefficients are zero-mean Gaussian distributed with variance 0.01. For the SMIG-NLMS and SIG-NLMS algorithms, the sliding data length is set as $L = 100$ and the kernel widths are chosen according to Silverman's rule. The step-sizes involved in the algorithms are experimentally selected so that the initial convergence rates are visually identical.

The average convergence curves of the angles $\theta(W_k, W^*)$ and $\theta(A_k, A^*)$, over 1000 independent Monte Carlo simulation runs, are shown in Figures 6.13 and 6.14. It is evident that the SMIG-NLMS algorithm achieves the smallest angles (mismatches) in both linear and nonlinear subsystems during the steady-state phase. More detailed statistical results of the subsystems training are presented in Figures 6.15 and 6.16, in which the histograms of the angles $\theta(W_k, W^*)$ and $\theta(A_k, A^*)$ at the final iteration are plotted. The inset plots in Figures 6.15 and 6.16 give the summary of the mean and spread of the histograms. One can observe again that the SMIG-NLMS algorithm outperforms both the LMS-NLMS and SIG-NLMS algorithms in terms of the angles between the estimated and true parameter vectors.

The overall identification performance can be measured by the IEP. Figure 6.17 illustrates the convergence curves of the IEP over 1000 Monte Carlo runs. It is clear that the SMIG-NLMS algorithm achieves the smallest IEP during the steady-state phase. Figure 6.18 shows the probability density functions of the steady-state

---

[7] In practice, the IEP is evaluated using the sample mean instead of the expectation value.

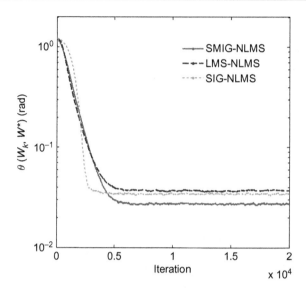

Figure 6.13 Average convergence curves of the angle $\theta(W_k, W^*)$ over 1000 Monte Carlo runs.
Source: Adopted from [136].

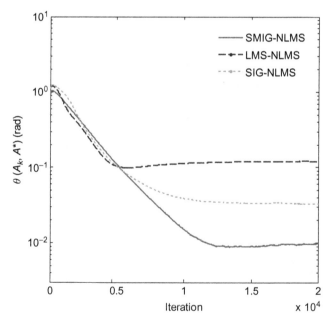

Figure 6.14 Average convergence curves of the angle $\theta(A_k, A^*)$ over 1000 Monte Carlo runs.
Source: Adopted from [136].

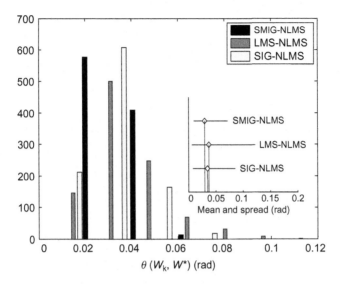

Figure 6.15 Histogram plots of the angle $\theta(W_k, W^*)$ at the final iteration over 1000 Monte Carlo runs.
Source: Adopted from [136].

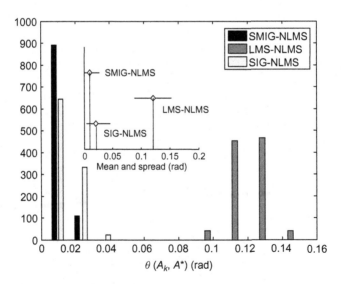

Figure 6.16 Histogram plots of the angle $\theta(A_k, A^*)$ at the final iteration over 1000 Monte Carlo runs.
Source: Adopted from [136].

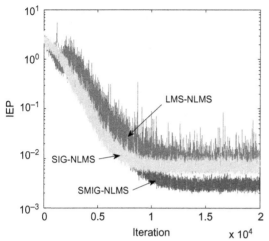

Figure 6.17 Convergence curves of the IEP over 1000 Monte Carlo runs.
Source: Adopted from [136].

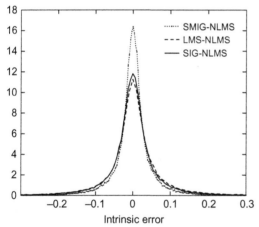

Figure 6.18 Probability density functions of the steady-state intrinsic errors.
Source: Adopted from [136].

intrinsic errors. As expected, the SMIG-NLMS algorithm yields the largest and most concentrated peak centered at the zero intrinsic error, and hence achieves the best accuracy in identification.

In the previous simulations, the unknown Wiener system has the same structure as the assumed model. In order to show how the algorithm performs when the real system is different from the assumed model (i.e., the unmatched case), another simulation with the same setup is conducted. This time the linear and nonlinear parts of the unknown system are assumed to be

$$\begin{cases} H(z) = \dfrac{1}{(1 - 0.3z^{-1})} \\ f(x) = \dfrac{1 - \exp(-x)}{1 + \exp(-x)} \end{cases} \tag{6.87}$$

Table 6.1 Mean ± Deviation Results of the IEP at the Final
Iteration Over 1000 Monte Carlo Runs

	IEP
SMIG-NLMS	0.0011 ± 0.0065
LMS-NLMS	0.0028 ± 0.0074
SIG-NLMS	0.0033 ± 0.0089

Source: Adopted from [136].

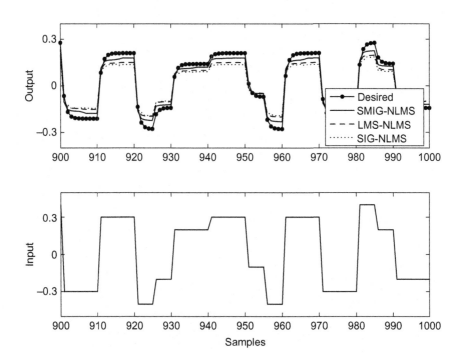

Figure 6.19 Desired output (the intrinsic output of the true system) and model outputs for the test input data.
Source: Adopted from [136].

Table 6.1 lists the mean ± deviation results of the IEP at final iteration over 1000 Monte Carlo runs. Clearly, the SMIG-NLMS algorithm produces the IEP with both lower mean and smaller deviation. Figure 6.19 shows the desired output (intrinsic output of the true system) and the model outputs (trained by different algorithms) during the last 100 samples for the test input. The results indicate that the identified model by the SMIG-NLMS algorithm describes the test data with the best accuracy.

Appendix I: MinMI Rate Criterion

The authors in [124] propose the MinMI rate criterion. Consider the linear Gaussian system

$$
\begin{cases}
\theta(t+1) = \theta(t) + v(t), \quad \theta(0) = \theta_0, \\
y(t) = \phi(t)^T \theta(t) + w(t)
\end{cases}
\tag{I.1}
$$

where $\theta_0 \sim \mathcal{N}(m_0, Q_0)$, $v(t) \sim \mathcal{N}(0, Q_v)$, $w(t) \sim \mathcal{N}(0, Q_w)$. One can adopt the following linear recursive algorithm to estimate the parameters

$$
\hat{\theta}(t+1) = \hat{\theta}(t) + K(t)[y(t) - \phi(t)^T \hat{\theta}(t)], \hat{\theta}(0) = m_0
\tag{I.2}
$$

The MinMI rate criterion is to search an optimal gain matrix $K^*(t)$ such that the mutual information rate $\bar{I}(\{e(t)\}; \{z(t)\})$ between the error signal $e(t) = \theta(t) - \hat{\theta}(t)$ and a unity-power white Gaussian noise $z(t)$ is minimized, where $z(t) = e(t) + r(t)$, $r(t)$ is a certain Gaussian process independent of $e(t)$. Clearly, the MinMI rate criterion requires that the asymptotic power spectral $R(z)$ of the error process $\{e(t)\}$ satisfies $0 \le R(z) \le I$ (otherwise $r(t)$ does not exist). It can be calculated that

$$
\begin{aligned}
\bar{I}(\{e(t)\}; \{z(t)\}) &= \frac{-1}{4\pi i} \int_D \frac{1}{z} \log \det[I - R(z)] \mathrm{d}z \\
&= \frac{-1}{4\pi i} \int_D \frac{1}{z} \log \det\left[I - S(z^{-1})^T S(z)\right] \mathrm{d}z
\end{aligned}
\tag{I.3}
$$

where $S(z)$ is a spectral factor of $R(z)$. Hence, under the MinMI rate criterion the optimal gain matrix K^* will be

$$
K^* = \arg \min_K \frac{-1}{4\pi i} \int_D \frac{1}{z} \log \det\left[I - S(z^{-1})^T S(z)\right] \mathrm{d}z
\tag{I.4}
$$

References

[1] L.A. Zadeh, From circuit theory to system theory, Proc. IRE 50 (5) (1962) 856–865.

[2] P. Eykhoff, System Identification—Parameter and State Estimation, John Wiley & Sons, Inc., London, 1974.

[3] L. Ljung, Convergence analysis of parametric identification methods, IEEE Trans. Automat. Control 23 (1978) 770–783.

[4] L. Ljung, System Identification: Theory for the User, second ed., Prentice Hall PTR, Upper Saddle River, New Jersey, 1999.

[5] P. Zhang, Model selection via multifold cross validation, Ann. Stat. (1993) 299–313.

[6] H. Akaike, Information theory and an extension of the maximum likelihood principle, Proceedings of the Second International Symposium on Information Theory, 1973, pp. 267–281.

[7] H. Akaike, A new look at the statistical model identification, IEEE Trans. Automat. Control 19 (6) (1974) 716–723.

[8] G. Schwarz, Estimating the dimension of a model, Ann. Stat. 6 (1978) 461–464.

[9] J. Rissanen, Modeling by shortest data description, Automatica 14 (1978) 465–471.

[10] A. Barron, J. Rissanen, B. Yu, The minimum description length principle in coding and modeling, IEEE Trans. Inf. Theory 44 (6) (1998) 2743–2760.

[11] C.R. Rojas, J.S. Welsh, G.C. Goodwin, A. Feuer, Robust optimal experiment design for system identification, Automatica 43 (6) (2007) 993–1008.

[12] W. Liu, J. Principe, S. Haykin, Kernel Adaptive Filtering: A Comprehensive Introduction, Wiley, 2010.

[13] J.R. Wolberg, J. Wolberg, Data Analysis Using the Method of Least Squares: Extracting the Most Information from Experiments, 1, Springer, Berlin, Germany, 2006.

[14] T. Kailath, A.H. Sayed, B. Hassibi, Linear Estimation, 1, Prentice Hall, New Jersey, 2000.

[15] J. Aldrich, R. A. Fisher and the making of maximum likelihood 1912–1922, Stat. Sci. 12 (3) (1997) 162–176.

[16] A. Hald, On the history of maximum likelihood in relation to inverse probability and least squares, Stat. Sci. 14 (2) (1999) 214–222.

[17] A.N. Tikhonov, V.Y. Arsenin, Solution of Ill-posed Problems, Winston & Sons, Washington, 1977.

[18] B. Widrow, S. Sterns, Adaptive Signal Processing, Prentice Hall, Englewood Cliffs, NJ, 1985.

[19] S. Haykin, Adaptive Filter Theory, Prentice Hall, Englewood Cliffs, NJ, 2002.

[20] S.S. Haykin, B. Widrow (Eds.), Least-Mean-Square Adaptive Filters, Wiley, New York, 2003.

[21] S. Sherman, Non-mean-square error criteria, IRE Trans. Inf. Theory 4 (1958) 125–126.

[22] J.L. Brown, Asymmetric non-mean-square error criteria, IRE Trans. Automat. Control 7 (1962) 64–66.

[23] M. Zakai, General error criteria, IEEE Trans. Inf. Theory 10 (1) (1964) 94–95.

[24] E.B. Hall, G.L. Wise, On optimal estimation with respect to a large family of cost function, IEEE Trans. Inf. Theory 37 (3) (1991) 691–693.

[25] L. Ljung, T. Soderstrom, Theory and Practice of Recursive Identification, MIT Press, Cambridge, MA, 1983.

[26] E. Walach, B. Widrow, The least mean fourth (LMF) adaptive algorithm and its family, IEEE. Trans. Inf. Theory 30 (2) (1984) 275–283.

[27] E. Walach, On high-order error criteria for system identification, IEEE Trans. Acoust. 33 (6) (1985) 1634–1635.

[28] S.C. Douglas, T.H.Y. Meng, Stochastic gradient adaptation under general error criteria, IEEE Trans. Signal Process. 42 (1994) 1335–1351.

[29] T.Y. Al-Naffouri, A.H. Sayed, Adaptive filters with error nonlinearities: mean-square analysis and optimum design, EURASIP J. Appl. Signal. Process. 4 (2001) 192–205.

[30] S.C. Pei, C.C. Tseng, Least mean p-power error criterion for adaptive FIR filter, IEEE J. Sel. Areas Commun. 12 (9) (1994) 1540–1547.

[31] M. Shao, C.L. Nikias, Signal processing with fractional lower order moments: stable processes and their applications, Proc. IEEE 81 (7) (1993) 986–1009.

[32] C.L. Nikias, M. Shao, Signal Processing with Alpha-Stable Distributions and Applications, Wiley, New York, 1995.

[33] P.J. Rousseeuw, A.M. Leroy, Robust Regression and Outlier Detection, John Wiley & Sons, Inc., New York, 1987.

[34] J.A. Chambers, O. Tanrikulu, A.G. Constantinides, Least mean mixed-norm adaptive filtering, Electron. Lett. 30 (19) (1994) 1574–1575.

[35] O. Tanrikulu, J.A. Chambers, Convergence and steady-state properties of the least-mean mixed-norm (LMMN) adaptive algorithm, IEE Proc. Vis. Image Signal Process. 143 (3) (1996) 137–142.

[36] J. Chambers, A. Avlonitis, A roust mixed-norm adaptive filter algorithm, IEEE Signal Process. Lett. 4 (2) (1997) 46–48.

[37] R.K. Boel, M.R. James, I.R. Petersen, Robustness and risk-sensitive filtering, IEEE Trans. Automat. Control 47 (3) (2002) 451–461.

[38] J.T. Lo, T. Wanner, Existence and uniqueness of risk-sensitive estimates, IEEE Trans. Automat. Control 47 (11) (2002) 1945–1948.

[39] A.N. Delopoulos, G.B. Giannakis, Strongly consistent identification algorithms and noise insensitive MSE criteria, IEEE Trans. Signal Process. 40 (8) (1992) 1955–1970.

[40] C.Y. Chi, W.T. Chen, Linear prediction based on higher order statistics by a new criterion, Proceedings of Sixth IEEE SP Workshop Stat. Array Processing, 1992.

[41] C.Y. Chi, W.J. Chang, C.C. Feng, A new algorithm for the design of linear prediction error filters using cumulant-based MSE criteria, IEEE Trans. Signal Process. 42 (10) (1994) 2876–2880.

[42] C.C. Feng, C.Y. Chi, Design of Wiener filters using a cumulant based MSE criterion, Signal Process. 54 (1996) 23–48.

[43] T.M. Cover, J.A. Thomas, Elements of Information Theory, John Wiley &Sons, Inc., Chichester, 1991.

[44] C.E. Shannon, A mathematical theory of communication, J. Bell Syst. Technol. 27 (379–423) (1948) 623–656.

[45] J.P. Burg, Maximum entropy spectral analysis, Proceedings of the Thirty-Seventh Annual International Social Exploration Geophysics Meeting, Oklahoma City, OK, 1967.

[46] M.A. Lagunas, M.E. Santamaria, A.R. Figueiras, ARMA model maximum entropy power spectral estimation, IEEE Trans. Acoust. 32 (1984) 984–990.

[47] S. Ihara, Maximum entropy spectral analysis and ARMA processes, IEEE Trans. Inf. Theory 30 (1984) 377−380.

[48] S.M. Kay, Modern Spectral Estimation: Theory and Application, Prentice Hall, Englewood Cliffs, NJ, 1988.

[49] R. Linsker, Self-organization in perceptual networks, Computer 21 (1988) 105−117.

[50] R. Linsker, How to generate ordered maps by maximizing the mutual information between input and output signals, Neural Comput. 1 (1989) 402−411.

[51] R. Linsker, Deriving receptive fields using an optimal encoding criterion, in: S.J. Hansor (Ed.), Proceedings of Advances in Neural Information Processing Systems, 1993, pp. 953−960.

[52] G. Deco, D. Obradovic, An Information—Theoretic Approach to Neural Computing, Springer-Verlag, New York, 1996.

[53] S. Haykin, Neural Networks: A Comprehensive Foundation, Prentice Hall, Inc., Englewood Cliffs, NJ, 1999.

[54] P. Comon, Independent component analysis, a new concept? Signal Process. 36 (3) (1994) 287−314.

[55] T.W. Lee, M. Girolami, T. Sejnowski, Independent component analysis using an extended infomax algorithm for mixed sub-Gaussian and super-Gaussian sources, Neural Comput. 11 (2) (1999) 409−433.

[56] T.W. Lee, M. Girolami, A.J. Bell, A unifying information-theoretic framework for independent component analysis, Comput. Math. Appl. 39 (11) (2000) 1−21.

[57] D. Erdogmus, K.E. Hild II, Y.N. Rao, J.C. Principe, Minimax mutual information approach for independent component analysis, Neural Comput. 16 (2004) 1235−1252.

[58] J.F. Cardoso, Infomax and maximum likelihood for blind source separation, IEEE Signal Process. Lett. 4 (1997) 109−111.

[59] H.H. Yang, S.I. Amari, Adaptive online learning algorithms for blind separation: maximum entropy minimum mutual information, Neural Comput. 9 (1997) 1457−1482.

[60] D.T. Pham, Mutual information approach to blind separation of stationary source, IEEE Trans. Inf. Theory 48 (7) (2002) 1−12.

[61] M.B. Zadeh, C. Jutten, A general approach for mutual information minimization and its application to blind source separation, Signal Process. 85 (2005) 975−995.

[62] J.C. Principe, D. Xu, J.W. Fisher, Information theoretic learning, in: S. Haykin (Ed.), Unsupervised Adaptive Filtering, Wiley, New York, 2000.

[63] J.C. Principe, D. Xu, Q. Zhao, et al., Learning from examples with information theoretic criteria, J. VLSI Signal Process. Syst. 26 (2000) 61−77.

[64] J.C. Principe, Information Theoretic Learning: Renyi's Entropy and Kernel Perspectives, Springer, New York, 2010.

[65] J.W. Fisher, Nonlinear Extensions to the Minimum Average Correlation Energy Filter, University of Florida, USA, 1997.

[66] D. Xu, Energy, *Entropy and Information Potential for Neural Computation*, University of Florida, USA, 1999.

[67] D. Erdogmus, Information Theoretic Learning: Renyi's Entropy and Its Applications to Adaptive System Training, University of Florida, USA, 2002.

[68] D. Erdogmus, J.C. Principe, From linear adaptive filtering to nonlinear information processing, IEEE Signal Process. Mag. 23 (6) (2006) 15−33.

[69] J. Zaborszky, An information theory viewpoint for the general identification problem, IEEE Trans. Automat. Control 11 (1) (1966) 130−131.

[70] H.L. Van Trees, Detection, Estimation, and Modulation Theory, Part I, John Wiley & Sons, New York, 1968.

[71] D.L. Snyder, I.B. Rhodes, Filtering and control performance bounds with implications on asymptotic separation, Automatica 8 (1972) 747−753.

[72] J.I.A. Galdos, Cramer−Rao bound for multidimensional discrete-time dynamical systems, IEEE Trans. Automat. Control 25 (1980) 117−119.

[73] B. Friedlander, J. Francos, On the accuracy of estimating the parameters of a regular stationary process, IEEE Trans. Inf. Theory 42 (4) (1996) 1202−1211.

[74] P. Stoica, T.L. Marzetta, Parameter estimation problems with singular information matrices, IEEE Trans. Signal Process. 49 (1) (2001) 87−90.

[75] L.P. Seidman, Performance limitations and error calculations for parameter estimation, Proc. IEEE 58 (1970) 644−652.

[76] M. Zakai, J. Ziv, Lower and upper bounds on the optimal filtering error of certain diffusion process, IEEE Trans. Inf. Theory 18 (3) (1972) 325−331.

[77] J.I. Galdos, A rate distortion theory lower bound on desired function filtering error, IEEE Trans. Inf. Theory 27 (1981) 366−368.

[78] R.B. Washburn, D. Teneketzis, Rate distortion lower bound for a special class of nonlinear estimation problems, Syst. Control Lett. 12 (1989) 281−286.

[79] H.L. Weidemann, E.B. Stear, Entropy analysis of estimating systems, IEEE Trans. Inf. Theory 16 (3) (1970) 264−270.

[80] T.E. Duncan, On the calculation of mutual information, SIAM J. Appl. Math. 19 (1970) 215−220.

[81] D. Guo, S. Shamai, S. Verdu, Mutual information and minimum mean-square error in Gaussian Channels, IEEE Trans. Inf. Theory 51 (4) (2005) 1261−1282.

[82] M. Zakai, On mutual information, likelihood-ratios and estimation error for the additive Gaussian channel, IEEE Trans. Inf. Theory 51 (9) (2005) 3017−3024.

[83] J. Binia, Divergence and minimum mean-square error in continuous-time additive white Gaussian noise channels, IEEE Trans. Inf. Theory 52 (3) (2006) 1160−1163.

[84] T.E. Duncan, B. Pasik-Duncan, Estimation and mutual information, Proceedings of the Forty-Sixth IEEE Conference on Decision and Control, New Orleans, LA, USA, 2007, pp. 324−327.

[85] S. Verdu, Mismatched estimation and relative entropy, IEEE Trans. Inf. Theory 56 (8) (2010) 3712−3720.

[86] H.L. Weidemann, E.B. Stear, Entropy analysis of parameter estimation, Inf. Control 14 (1969) 493−506.

[87] Y. Tomita, S. Ohmatsu, T. Soeda, An application of the information theory to estimation problems, Inf. Control 32 (1976) 101−111.

[88] P. Kalata, R. Priemer, Linear prediction, filtering, and smoothing: an information theoretic approach, Inf. Sci. (Ny) 17 (1979) 1−14.

[89] N. Minamide, An extension of the entropy theorem for parameter estimation, Inf. Control 53 (1) (1982) 81−90.

[90] M. Janzura, T. Koski, A. Otahal, Minimum entropy of error principle in estimation, Inf. Sci. (Ny). 79 (1994) 123−144.

[91] T.L. Chen, S. Geman, On the minimum entropy of a mixture of unimodal and symmetric distributions, IEEE Trans. Inf. Theory 54 (7) (2008) 3166−3174.

[92] B. Chen, Y. Zhu, J. Hu, M. Zhang, On optimal estimations with minimum error entropy criterion, J. Franklin Inst. 347 (2) (2010) 545−558.

[93] B. Chen, Y. Zhu, J. Hu, M. Zhang, A new interpretation on the MMSE as a robust MEE criterion, Signal Process. 90 (12) (2010) 3313−3316.

[94] B. Chen, J.C. Principe, Some further results on the minimum error entropy estimation, Entropy 14 (5) (2012) 966−977.

[95]   B. Chen, J.C. Principe, On the smoothed minimum error entropy criterion, Entropy 14 (11) (2012) 2311−2323.

[96]   E. Parzen, On estimation of a probability density function and mode, Time Series Analysis Papers, Holden-Day, Inc., San Diego, CA, 1967.

[97]   B.W. Silverman, Density Estimation for Statistic and Data Analysis, Chapman & Hall, NY, 1986.

[98]   L. Devroye, G. Lugosi, Combinatorial Methods in Density Estimation, Springer-Verlag, New York, 2000.

[99]   I. Santamaria, D. Erdogmus, J.C. Principe, Entropy minimization for supervised digital communications channel equalization, *IEEE Trans.* Signal Process. 50 (5) (2002) 1184−1192.

[100] D. Erdogmus, J.C. Principe, An error-entropy minimization algorithm for supervised training of nonlinear adaptive systems, *IEEE Trans.* Signal Process. 50 (7) (2002) 1780−1786.

[101] D. Erdogmus, J.C. Principe, Generalized information potential criterion for adaptive system training, IEEE Trans. Neural Netw. 13 (2002) 1035−1044.

[102] D. Erdogmus, J.C. Principe, Convergence properties and data efficiency of the minimum error entropy criterion in Adaline training, *IEEE Trans.* Signal Process. 51 (2003) 1966−1978.

[103] D. Erdogmus, K.E. Hild II, J.C. Principe, Online entropy manipulation: stochastic information gradient, *IEEE* Signal Process. *Lett.* 10 (2003) 242−245.

[104] R.A. Morejon, J.C. Principe, Advanced search algorithms for information-theoretic learning with kernel-based estimators, IEEE Trans. Neural Netw. 15 (4) (2004) 874−884.

[105] S. Han, S. Rao, D. Erdogmus, K.H. Jeong, J.C. Principe, A minimum-error entropy criterion with self-adjusting step-size (MEE-SAS), Signal Process. 87 (2007) 2733−2745.

[106] B. Chen, Y. Zhu, J. Hu, Mean-square convergence analysis of ADALINE training with minimum error entropy criterion, IEEE Trans. Neural Netw. 21 (7) (2010) 1168−1179.

[107] L.Z. Guo, S.A. Billings, D.Q. Zhu, An extended orthogonal forward regression algorithm for system identification using entropy, Int. J. Control. 81 (4) (2008) 690−699.

[108] S. Kullback, Information Theory and Statistics, John Wiley & Sons, New York, 1959.

[109] R.A. Kulhavý, Kullback−Leibler distance approach to system identification, Annu. Rev. Control 20 (1996) 119−130.

[110] T. Matsuoka, T.J. Ulrych, Information theory measures with application to model identification, IEEE. Trans. Acoust. 34 (3) (1986) 511−517.

[111] J.E. Cavanaugh, A large-sample model selection criterion based on Kullback's symmetric divergence, Stat. Probab. Lett. 42 (1999) 333−343.

[112] A.K. Seghouane, M. Bekara, A small sample model selection criterion based on the Kullback symmetric divergence, *IEEE Trans.* Signal Process. 52 (12) (2004) 3314−3323.

[113] A.K. Seghouane, S.I. Amari, The AIC criterion and symmetrizing the Kullback−Leibler divergence, IEEE Trans. Neural Netw. 18 (1) (2007) 97−106.

[114] A.K. Seghouane, Asymptotic bootstrap corrections of AIC for linear regression models, Signal Process. 90 (1) (2010) 217−224.

[115] Y. Baram, N.R. Sandell, An information theoretic approach to dynamic systems modeling and identification, IEEE Trans. Automat. Control 23 (1) (1978) 61−66.

[116] Y. Baram, Y. Beeri, Stochastic model simplification, IEEE Trans. Automat. Control 26 (2) (1981) 379−390.

[117] J.K. Tugnait, Continuous-time stochastic model simplification, IEEE Trans. Automat. Control 27 (4) (1982) 993−996.

[118] R. Leland, Reduced-order models and controllers for continuous-time stochastic systems: an information theory approach, IEEE Trans. Automat. Control 44 (9) (1999) 1714−1719.

[119] R. Leland, An approximate-predictor approach to reduced-order models and controllers for distributed-parameter systems, IEEE Trans. Automat. Control 44 (3) (1999) 623−627.

[120] B. Chen, J. Hu, Y. Zhu, Z. Sun, Parameter identifiability with Kullback−Leibler information divergence criterion, Int. J. Adapt. Control Signal Process. 23 (10) (2009) 940−960.

[121] E. Weinstein, M. Feder, A.V. Oppenheim, Sequential algorithms for parameter estimation based on the Kullback−Leibler information measure, IEEE. Trans. Acoust. 38 (9) (1990) 1652−1654.

[122] V. Krishnamurthy, Online estimation of dynamic shock-error models based on the Kullback−Leibler information measure, IEEE Trans. Automat. Control 39 (5) (1994) 1129−1135.

[123] A.A. Stoorvogel, J.H. Van Schuppen, Approximation problems with the divergence criterion for Gaussian variables and Gaussian process, Syst. Control. Lett. 35 (1998) 207−218.

[124] A.A. Stoorvogel, J.H. Van Schuppen, System identification with information theoretic criteria, in: S. Bittanti, G. Picc (Eds.), Identification, Adaptation, Learning, Springer, Berlin, 1996.

[125] L. Pu, J. Hu, B. Chen, Information theoretical approach to identification of hybrid systems, Hybrid Systems: Computation and Control, Springer, Berlin Heidelberg, 2008, pp. 650−653

[126] B. Chen, Y. Zhu, J. Hu, Z. Sun, Adaptive filtering under minimum information divergence criterion, Int. J. Control Autom. Syst. 7 (2) (2009) 157−164.

[127] S.A. Chandra, M. Taniguchi, Minimum α-divergence estimation for ARCH models, J. Time Series Anal. 27 (1) (2006) 19−39.

[128] M.C. Pardo, Estimation of parameters for a mixture of normal distributions on the basis of the Cressie and Read divergence, Commun. Stat-Simul. C. 28 (1) (1999) 115−130.

[129] N. Cressie, L. Pardo, Minimum ϕ-divergence estimator and hierarchical testing in loglinear models, Stat. Sin. 10 (2000) 867−884.

[130] L. Pardo, Statistical Inference Based on Divergence Measures, Chapman & Hall/CRC, Boca Raton, FL, 2006.

[131] X. Feng, K.A. Loparo, Y. Fang, Optimal state estimation for stochastic systems: an information theoretic approach, IEEE Trans. Automat. Control 42 (6) (1997) 771−785.

[132] D. Mustafa, K. Glover, Minimum entropy H_∞ control, Lecture Notes in Control and Information Sciences, 146, Springer-Verlag, Berlin, 1990.

[133] J.-M. Yang, H. Sakai, A robust ICA-based adaptive filter algorithm for system identification, IEEE Trans. Circuits Syst. Express Briefs 55 (12) (2008) 1259−1263.

[134] I.S. Durgaryan, F.F. Pashchenko, Identification of objects by the maximal information criterion, Autom. Remote Control 62 (7) (2001) 1104−1114.

[135] B. Chen, J. Hu, H. Li, Z. Sun, Adaptive filtering under maximum mutual information criterion, Neurocomputing 71 (16) (2008) 3680−3684.

[136] B. Chen, Y. Zhu, J. Hu, J.C. Príncipe, Stochastic gradient identification of Wiener system with maximum mutual information criterion, Signal Process., IET 5 (6) (2011) 589−597.

[137] W. Liu, P.P. Pokharel, J.C. Principe, Correntropy: properties and applications in non-Gaussian signal processing, *IEEE Trans.* Signal Process. 55 (11) (2007) 5286−5298.

[138] A. Singh, J.C. Principe, Using correntropy as a cost function in linear adaptive filters, in: International Joint Conference on Neural Networks (IJCNN'09), IEEE, 2009, pp. 2950−2955.

[139] S. Zhao, B. Chen, J.C. Principe, Kernel adaptive filtering with maximum correntropy criterion, in: The 2011 International Joint Conference on Neural Networks (IJCNN), IEEE, 2011, pp. 2012−2017.

[140] R.J. Bessa, V. Miranda, J. Gama, Entropy and correntropy against minimum square error in offline and online three-day ahead wind power forecasting, IEEE Trans. Power Systems 24 (4) (2009) 1657−1666.

[141] Xu J.W., Erdogmus D., Principe J.C. Minimizing Fisher information of the error in supervised adaptive filter training, in: *Proc. ICASSP*, 2004, pp.513−516.

[142] Y. Sakamoto, M. Ishiguro, G. Kitagawa, Akaike Information Criterion Statistics, Reidel Publishing Company, Dordretcht, Netherlands, 1986.

[143] K.P. Burnham, D.R. Anderson, Model Selection and Multimodel Inference: A Practical Information Theoretic Approach, second ed., Springer-Verlag, New York, 2002.

[144] P.D. Grunwald, The Minimum Description Length Principle, MIT Press, Cambridge, MA, 2007.

[145] N. Tishby, F.C. Pereira, W. Bialek, The information bottleneck method. arXiv preprint physics/0004057, 2000.

[146] A.N. Kolmogorov, Three approaches to the quantitative definition of information, Probl. Inform. Transm. 1 (1965) 4−7.

[147] O. Johnson, O.T. Johnson, Information Theory and the Central Limit Theorem, Imperial College Press, London, 2004.

[148] E.T. Jaynes, Information theory and statistical mechanics, Phys. Rev. 106 (1957) 620−630.

[149] J.N. Kapur, H.K. Kesavan, Entropy Optimization Principles with Applications, Academic Press, Inc., 1992.

[150] D. Ormoneit, H. White, An efficient algorithm to compute maximum entropy densities, Econom. Rev. 18 (2) (1999) 127−140.

[151] X. Wu, Calculation of maximum entropy densities with application to income distribution, J. Econom. 115 (2) (2003) 347−354.

[152] A. Renyi, On measures of entropy and information. Proceedings of the Fourth Berkeley Symposium on Mathematical Statistics and Probability, vol. 1, 1961, pp. 547−561.

[153] J. Havrda, F. Charvat, Concept of structural α-entropy, Kybernetika 3 (1967) 30−35.

[154] R.S. Varma, Generalizations of Renyi's entropy of order α, J. Math. Sci. 1 (1966) 34−48.

[155] S. Arimoto, Information-theoretic considerations on estimation problems, Inf. Control 19 (1971) 181−194.

[156] M. Salicu, M.L. Menendez, D. Morales, L. Pardo, Asymptotic distribution of (h, ϕ)-entropies, Commun. Stat. Theory Methods 22 (1993) 2015−2031.

[157] M. Rao, Y. Chen, B.C. Vemuri, F. Wang, Cumulative residual entropy: a new measure of information, IEEE Trans. Inf. Theory 50 (6) (2004) 1220−1228.

[158] K. Zografos, S. Nadarajah, Survival exponential entropies, IEEE Trans. Inf. Theory 51 (3) (2005) 1239−1246.

[159] B. Chen, P. Zhu, J.C. Príncipe, Survival information potential: a new criterion for adaptive system training, *IEEE Trans.* Signal Process. 60 (3) (2012) 1184−1194.

[160] J. Mercer, Functions of positive and negative type, and their connection with the theory of integral equations, Philos. Trans. R. Soc. London 209 (1909) 415−446.

[161] V. Vapnik, The Nature of Statistical Learning Theory, Springer, New York, 1995.

[162] B. Scholkopf, A.J. Smola, Learning with Kernels, Support Vector Machines, Regularization, Optimization and Beyond, MIT Press, Cambridge, MA, USA, 2002.

[163] P. Whittle, The analysis of multiple stationary time series, J. R. Stat. Soc. B 15 (1) (1953) 125−139.

[164] A.P. Dempster, N.M. Laird, D.B. Rubin, Maximum likelihood from incomplete data via the EM algorithm, J. R. Stat. Soc. B 39 (1) (1977) 1−38.

[165] G.J. McLachlan, T. Krishnan, The EM Algorithm and Extensions, Wiley-Interscience, NJ, 2008.

[166] B. Aiazzi, L. Alparone, S. Baronti, Estimation based on entropy matching for generalized Gaussian PDF modeling, *IEEE* Signal Process. *Lett.* 6 (6) (1999) 138−140.

[167] D.T. Pham, Entropy of a variable slightly contaminated with another, *IEEE* Signal Process. *Lett.* 12 (2005) 536−539.

[168] B. Chen, J. Hu, Y. Zhu, Z. Sun, Information theoretic interpretation of error criteria, Acta Automatica Sin. 35 (10) (2009) 1302−1309.

[169] B. Chen, J.C. Príncipe, Maximum correntropy estimation is a smoothed MAP estimation, *IEEE* Signal Process. *Lett.* 19 (8) (2012) 491−494.

[170] R.Y. Rubinstein, Simulation and the Monte Carlo Method, Wiley, New York, 1981.

[171] M.A. Styblinsli, T.S. Tang, Experiments in nonconvex optimization: stochastic approximation with function smoothing and simulated annealing, Neural Netw. 3 (1990) 467−483.

[172] W. Edmonson, K. Srinivasan, C. Wang, J. Principe, A global least mean square algorithm for adaptive IIR filtering, IEEE Trans. Circuits Syst 45 (1998) 379−384.

[173] G.C. Goodwin, R.L. Payne, Dynamic System Identification: Experiment Design and Data Analysis, Academic Press, New York, 1977.

[174] E. Moore, On properly positive Hermitian matrices, Bull. Amer. Math. Soc. 23 (59) (1916) 66−67.

[175] N. Aronszajn, The theory of reproducing kernels and their applications, Cambridge Philos. Soc. Proc. 39 (1943) 133−153.

[176] Y. Engel, S. Mannor, R. Meir, The kernel recursive least-squares algorithm, *IEEE Trans.* Signal Process. 52 (2004) 2275−2285.

[177] W. Liu, P. Pokharel, J. Principe, The kernel least mean square algorithm, *IEEE Trans.* Signal Process. 56 (2008) 543−554.

[178] W. Liu, J. Principe, Kernel affine projection algorithm, EURASIP J. Adv. Signal Process. (2008)10.1155/2008/784292Article ID 784292, 12 pages

[179] J. Platt, A resource-allocating network for function interpolation, Neural. Comput. 3 (1991) 213−225.

[180] C. Richard, J.C.M. Bermudez, P. Honeine, Online prediction of time series data with kernels, *IEEE Trans.* Signal Process. 57 (2009) 1058−1066.

[181] W. Liu, I.l. Park, J.C. Principe, An information theoretic approach of designing sparse kernel adaptive filters, IEEE Trans. Neural Netw. 20 (2009) 1950−1961.

[182] B. Chen, S. Zhao, P. Zhu, J.C. Principe, Quantized kernel least mean square algorithm, IEEE Trans. Neural Netw. Learn. Syst. 23 (1) (2012) 22−32.

[183] J. Berilant, E.J. Dudewicz, L. Gyorfi, E.C. van der Meulen, Nonparametric entropy estimation: an overview, Int. J. Math. Statist. Sci. 6 (1) (1997) 17−39.

[184] O. Vasicek, A test for normality based on sample entropy, J. Roy. Statist. Soc. B 38 (1) (1976) 54−59.

[185] A. Singh, J.C. Príncipe, Information theoretic learning with adaptive kernels, Signal Process. 91 (2) (2011) 203−213.

[186] D. Erdogmus, J.C. Principe, S.-P. Kim, J.C. Sanchez, A recursive Renyi's entropy estimator, in: Proceedings of the Twelfth IEEE Workshop on Neural Networks for Signal Processing, 2002, pp. 209−217.

[187] X. Wu, T. Stengos, Partially adaptive estimation via the maximum entropy densities, J. Econom. 8 (2005) 352−366.

[188] B. Chen, Y. Zhu, J. Hu, M. Zhang, Stochastic information gradient algorithm based on maximum entropy density estimation, ICIC Exp. Lett. 4 (3) (2010) 1141−1145.

[189] Y. Zhu, B. Chen, J. Hu, Adaptive filtering with adaptive p-power error criterion, Int. J. Innov. Comput. Inf. Control 7 (4) (2011) 1725−1738.

[190] B. Chen, J.C. Principe, J. Hu, Y. Zhu, Stochastic information gradient algorithm with generalized Gaussian distribution model, J. Circuit. Syst. Comput. 21 (1) (2012).

[191] M.K. Varanasi, B. Aazhang, Parametric generalized Gaussian density estimation, J. Acoust. Soc. Amer. 86 (4) (1989) 1404−1415.

[192] K. Kokkinakis, A.K. Nandi, Exponent parameter estimation for generalized Gaussian probability density functions with application to speech modeling, Signal Process. 85 (2005) 1852−1858.

[193] S. Han, J.C. Principe, A fixed-point minimum error entropy algorithm, in: Proceedings of the Sixteenth IEEE Signal Processing Society Workshop on Machine Learning for Signal Processing, 2006, pp.167−172.

[194] S. Han, A family of minimum Renyi's error entropy algorithm for information processing, Doctoral dissertation, University of Florida, 2007.

[195] S. Chen, S.A. Billings, P.M. Grant, Recursive hybrid algorithm for non-linear system identification using radial basis function networks, Int. J. Control. 55 (1992) 1051−1070.

[196] A. Sayed, Fundamentals of Adaptive Filtering, Wiley, New York, 2003.

[197] G.A. Clark, S.K. Mitra, S.R. Parker, Block implementation of adaptive digital filters, IEEE Trans. Acoust. Speech Signal Process. ASSP-29 (3) (1981) 744−752.

[198] N.J. Bershad, M. Bonnet, Saturation effects in LMS adaptive echo cancellation for binary data, IEEE Trans. Acoust. Speech Signal Process 38 (10) (1990) 1687−1696.

[199] T.Y. Al-Naffouri, A. Zerguine, M. Bettayeb, Convergence analysis of the LMS algorithm with a general error nonlinearity and an iid input, in: Proceedings of the Asilomar Conference on Signals, Systems, and Computers, vol. 1, 1998, pp. 556−559.

[200] B. Chen, J. Hu, L. Pu, Z. Sun, Stochastic gradient algorithm under (h,φ)-entropy criterion, *Circuits Syst.* Signal Process. 26 (6) (2007) 941−960.

[201] J.D. Gibson, S.D. Gray, MVSE adaptive filtering subject to a constraint on MSE, IEEE Trans. Circuits Syst. 35 (5) (1988) 603−608.

[202] B.L.S.P. Rao, Asymptotic Theory of Statistical Inference, Wiley, New York, 1987.

[203] D. Kaplan, L. Glass, Understanding Nonlinear Dynamics, Springer-Verlag, New York, 1995.

[204] J.M. Kuo, Nonlinear dynamic modeling with artificial neural networks, Ph.D. dissertation, University of Florida, Gainesville, 1993

[205] D.G. Luenberger, Linear and Nonlinear Programming, Addison-Wesley, Reading, MA, 1973.

[206] L.Y. Wang, J.F. Zhang, G.G. Yin, System identification using binary sensors, IEEE Trans. Automat. Control 48 (11) (2003) 1892−1907.

[207] A.C. Harvey, C. Fernandez, Time series for count data or qualitative observations, J. Bus. Econ. Stat. 7 (1989) 407−417.

[208] M. Al-Osh, A. Alzaid, First order integer-valued autoregressive INAR(1) process, J. Time Series Anal. 8 (3) (1987) 261–275.

[209] K. Brannas, A. Hall, Estimation in integer-valued moving average models, Appl. Stoch. Model. Bus. Ind. 17 (3) (2001) 277–291.

[210] C.H. Weis, Thinning operations for modeling time series of counts—a survey, AStA Adv. Stat. Anal. 92 (3) (2008) 319–341.

[211] B. Chen, Y. Zhu, J. Hu, J.C. Principe, Δ-Entropy: definition, properties and applications in system identification with quantized data, Inf. Sci. 181 (7) (2011) 1384–1402.

[212] M. Janzura, T. Koski, A. Otahal, Minimum entropy of error estimation for discrete random variables, IEEE Trans. Inf. Theory 42 (4) (1996) 1193–1201.

[213] L.M. Silva, C.S. Felgueiras, L.A. Alexandre, J. Marques, Error entropy in classification problems: a univariate data analysis, Neural Comput. 18 (2006) 2036–2061.

[214] U. Ozertem, I. Uysal, D. Erdogmus, Continuously differentiable sample-spacing entropy estimation, IEEE Trans. Neural Netw. 19 (2008) 1978–1984.

[215] P. Larranaga, J.A. Lozano, Estimation of Distribution Algorithms: A New Tool for Evolutionary Computation, Kluwer Academic Publishers, Boston, 2002.

[216] T.J. Rothenberg, Identification in parametric models, Econometrica 39 (1971) 577–591.

[217] M.S. Grewal, K. Glover, Identifiability of linear and nonlinear dynamic systems, IEEE Trans. Automat. Control 21 (6) (1976) 833–837.

[218] E. Tse, J.J. Anton, On the identifiability of parameters, IEEE Trans. Automat. Control 17 (5) (1972) 637–646.

[219] K. Glover, J.C. Willems, Parameterizations of linear dynamical systems: canonical forms and identifiability, IEEE Trans. Automat. Control 19 (6) (1974) 640–646.

[220] A.W. van der Vaart, Asymptotic Statistics, Cambridge Series in Statistical and Probabilistic Mathematics, Cambridge University Press, New York, 1998.

[221] J.L. Doob, Stochastic Processes, John Wiley, New York, 1953.

[222] A. Sara, van de Geer, Empirical Processes in M-estimation, Cambridge Series in Statistical and Probabilistic Mathematics, Cambridge University Press, Cambridge, 2000.

[223] A.R. Barron, L. Gyorfi, E.C. van der Meulen, Distribution estimation consistent in total variation and in two types of information divergence, IEEE Trans. Inf. Theory 38 (5) (1992) 1437–1454.

[224] A. Papoulis, S.U. Pillai, Probability, Random Variables, and Stochastic Processes, fourth ed., McGraw-Hill Companies, Inc., New York, 2002.

[225] M. Karny, Towards fully probabilistic control design, Automatica 32 (12) (1996) 1719–1722.

[226] M. Karny, T.V. Guy, Fully probabilistic control design, Syst. Control Lett. 55 (2006) 259–265.

[227] H. Wang, Robust control of the output probability density functions for multivariable stochastic systems with guaranteed stability, IEEE Trans. Automat. Control 44 (11) (1999) 2103–2107.

[228] H. Wang, Bounded Dynamic Stochastic Systems: Modeling and Control, Springer-Verlag, New York, 2000.

[229] H. Wang, H. Yue, A rational spline model approximation and control of output probability density functions for dynamic stochastic systems, Trans. Inst. Meas. Control 25 (2) (2003) 93–105.

[230] J. Sala-Alvarez, G. Vázquez-Grau, Statistical reference criteria for adaptive signal processing in digital communications, *IEEE Trans.* Signal Process. 45 (1) (1997) 14–31.

[231] M.E. Meyer, D.V. Gokhale, Kullback—Leibler information measure for studying convergence rates of densities and distributions, IEEE Trans. Inf. Theory 39 (4) (1993) 1401—1404.

[232] R. Vidal, B. Anderson, Recursive identification of switched ARX hybrid models: exponential convergence and persistence of excitation, Proceedings of the Forty-Third IEEE Conference on Decision and Control (CDC), 2004.

[233] C.-A. Lai, Global optimization algorithms for adaptive infinite impulse response filters, Ph.D. Dissertation, University of Florida, 2002.

[234] B. Chen, J. Hu, H. Li, Z. Sun, Adaptive FIR filtering under minimum error/input information criterion. The Seventeenth IFAC Word Conference, Seoul, Korea, July 2008, pp. 3539—3543.

[235] T. Kailath, B. Hassibi, Linear Estimation, Prentice Hall, NJ, 2000.

[236] A. Hyvarinen, J. Karhunen, E. Oja, Independent Component Analysis, Wiley, New York, 2001.

[237] J.F. Cardoso, B.H. Laheld, Equivariant adaptive source separation, *IEEE Trans. Signal Process.* 44 (12) (1996) 3017—3030.

[238] M. Schetzen, The Volterra and Wiener Theories of Nonlinear Systems, Wiley, New York, 1980.

[239] N.J. Bershad, P. Celka, J.M. Vesin, Stochastic analysis of gradient adaptive identification of nonlinear systems with memory for Gaussian data and noisy input and output measurements, *IEEE Trans.* Signal Process. 47 (1999) 675—689.

[240] P. Celka, N.J. Bershad, J.M. Vesin, Stochastic gradient identification of polynomial Wiener systems: analysis and application, *IEEE Trans.* Signal Process. 49 (2001) 301—313.

Printed in the United States
By Bookmasters